What's so Special about Biodynamic Wine?

What's so Special about Biodynamic Wine?

Antoine Lepetit de la Bigne

Floris Books

Whenever you find yourself on the side of the majority, it is time to pause and reflect.

Mark Twain

To Christine

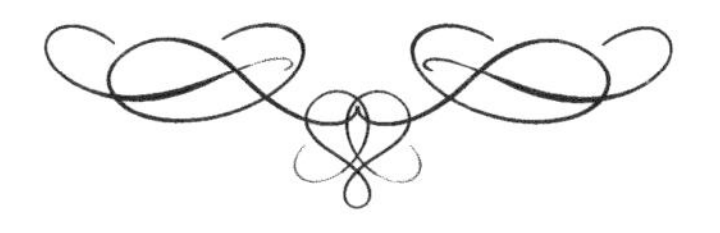

My sincere thanks to those who have guided me along the paths of wine, tasting and viticulture: Philippe Bourguignon, Olivier Humbrecht, Anne-Claude Leflaive, Bruno Quenioux, and the others.

Published in 2013 by Floris Books

Antoine Lepetit de la Bigne has asserted his right under the Copyright, Designs and Patents Act 1988 to be identified as the Author of this Work

This book is also available as an eBook

British Library CIP data available
ISBN 978-178250-021-6
Printed in Great Britain
by Bell & Bain Ltd, Glasgow

Contents

Foreword

One sentence of this introduction to biodynamic wine by Antoine Lepetit-de la Bigne remains imprinted in my mind because I believe it is essential to define the new relation of people with wine: 'In our post-industrial, overly-urbanised society, could wine be our last strong tie to the earth?' Our civilisation, nobody will deny, tends to eradicate the old traditional relation of man with nature. Towns grow irrevocably, the countryside becomes a desert. As for agricultural products, they become more and more industrialised, technologically crafted, less and less the result of the ancestral marriage of human beings with the earth. Wine also can be technologically crafted and a lot of wines are made that way. The wine connoisseurs of our time feel deeply this loss of the old tradition, and this is why for them the quality of a wine is profoundly linked to its origin, to the earth it comes from. Is the wine I buy a good representation of the place it comes from? Has the *vigneron,* like a good 'midwife', been able to establish with their vineyards this living relationship that will allow their qualities and their character to be expressed at their potential best?

In their search for the best expression of a wine's origin, *vignerons* inevitably come across the organic options, normal organic viticulture and biodynamics, since these methods aim at creating conditions where the *terroir,* that is, the soil and its environment, will emerge at their best.

Biodynamics, which deals essentially with what is not quantifiable – 'life forces' which science cannot measure, weigh or analyse – represents a challenge in making itself understood by the public. Until recently it used to be mocked by scientists. A lot of rumours sprang up regarding dynamisation, the various preparations, cosmic influences ... This book is essential because, with simplicity and humility, it first answers those legitimate questions raised by a method that both fascinates and troubles the public, and second because it explains why the *vignerons* – among them some of the most advanced of the industry – find in biodynamics that link to the earth which is so essential for delivering the potential of *terroirs* which have been cared for by human hand for centuries. As with biodynamics now, so it was before our industrialised times; that is to say: 'Nature is an open book for those who cultivate the ground with their hands.'

Who could be in a better place to write this book than Antoine Lepetit de la Bigne? The reasons are multiple: Antoine arrived in Burgundy a few years ago when he was very young, and my experience has shown me on many occasions how much, for a company or an industry, the arrival of new vision is important and enriching. Furthermore, having brilliantly achieved degree level in Science at the prestigious École Polytechnique, Antoine successfully moved across to graduate in agronomical studies. From there, always guided by his questing and free spirit, he went into viticulture practice with a Domaine which is at the forefront of this biodynamics that science so often denigrates!

Antoine has absorbed all that science has taught him and he is now like many scientists who have spent their life studying methodically the vine and the wine and who, after these years of accumulated scientific knowledge, know that, in the making of a great wine, what you don't understand is more important than what you understand or can measure or analyse.

Intuition, observation, humility and, last but not least, rigour, become the key qualities of the *vigneron.* Through Antoine's patient, careful and precise explanations, this book introduces us all, wine lovers, professionals, wine makers, sommeliers and others from the world of wine, to the fundamental lessons taught by biodynamics. The 'small voice' of biodynamics has found its ambassador.

Aubert de Villaine, Domaine de la Romanée-Conti

Preface

Dear reader and wine lover

Have you ever found yourself listening to a producer or wine merchant who raved on about the qualities of an organic or biodynamic wine using language such as: 'natural', 'mineral', 'grown with energy lines and cosmic forces', 'harvested with the moon' ...?

Maybe you thought this sounded so great that you rushed out to buy the wine. Maybe you experienced strange sensations when you tasted it. Maybe, even, you slept better after drinking it, without any unpleasant surprises – such as a headache – the next morning.

Or, as poetic as all this talk may seem, did it come across as rather obscure, trite, downright confusing, and not really respectful of all the scientific advances of the last two centuries?

In short, maybe you decided it was just the latest sales talk for describing even more qualities of *terroir.*

Over the last few years, when talking with clients during tastings at Domaine Leflaive and at the Puligny-Montrachet *École du Vin et des Terroirs,* I have come to realise how much confusion surrounds biodynamics. Although this term has become quite fashionable in the wine world, few wine enthusiasts have a clear understanding of what biodynamics really is. Some have a completely false impression, and others have no idea at all.

The term *biodynamic* resonates strongly with our subconscious, and generally exposes a range of preconceived ideas, especially in our Western culture, and particularly in France.

It is for this reason that I wanted to write this book. I hope to provide you with an introduction to biodynamics which is brief and user-friendly in both its form and content, and which responds to the questions you might ask yourself.

It is not my intention to explore in-depth the philosophy of Rudolf Steiner, and even less to offer a grower's handbook on biodynamic viticulture. Excellent works on this subject exist and are referenced in the bibliography (François Bouchet, Nicolas Joly, and Pierre Masson among others).

I have used clear language and sought to call on my knowledge of both science and the practice of biodynamic viticulture. The book reflects my current viewpoint and I assume full responsibility for it.

I want to thank the numerous wine enthusiasts around the world who have asked me the questions posed here, and who made me aware of the need for such a simple, comprehensible book. Although simplistic at times, it is intended to be a first step toward dispelling illusions about biodynamics. Afterwards, it is up to you to deepen your knowledge of the subject.

Antoine Lepetit de la Bigne

1 *Are biodynamic wines better than other wines?*

Ah, the big question that every wine enthusiast inevitably asks in their pursuit of perfection! To those looking for an immediate answer, I must say that I will avoid responding in a clear-cut manner at this, the beginning of the book. Dear reader, along the path I propose to follow in the world of biodynamics, you will form your own response to this question. It will become evident.

Given that biodynamic viticultural practices are more in harmony with the laws of nature (see question 11: In what way is biodynamics more respectful of nature?), I would, of course, be tempted to say that the resulting grape (the raw material from which wine is made) has greater potential. And then? A winegrower's skill also applies to the work in the cellar, the place where the potential is revealed ... or not (see question 16: In the winery: does biodynamics apply to vinification?). And what can be said about the concept of 'better wine'? In today's world we are inundated with wine rankings and scores. These scores inevitably simplify each wine, yet are immensely reassuring for wine lovers who cannot taste the wines themselves, or formulate their own opinion. How many 'experts' (professional or amateur) contribute to this with their influential opinions and, of course, the commercial and financial consequences for producers! The American wine critic, Robert Parker, was certainly the first, or in any case, the most emblematic in this trend. Today, however, every wine lover must break free from this by subjectively rediscovering and taking responsibility for their own likes and dislikes. It seems to me that in this respect, the increasing role assumed by women in the wine world, from growers to consumers, will be the primary catalyst for this necessary change (see question 20: How should a biodynamic wine be tasted?).

At this stage I will simply make two observations: the first, of a general nature, is that more and more top producers recognised for their pursuit of excellence in Burgundy, Alsace and other regions of France and the world, are opting for biodynamic viticulture.

The second, of a personal nature, is that I find myself essentially bringing wines into my cellar only from producers I know personally, most of whom practise biodynamic viticulture. Being curious by nature, I used to be very eclectic in my choices. These days, however, I have a number of bottles (some with very high scores from Mr Parker!) that I can no longer drink. Strangely enough, I never seem to find the opportunity to drink them. Could there be another dimension of quality in wine linked to the man or woman who produced it, and to the form of viticulture in which it was produced?

Part 1

VITICULTURE

2 *Why is biodynamics fashionable today?*

This form of agriculture was established more than eighty years ago and is based on lectures given by Rudolf Steiner in 1924 (see question 31: What is the content of the Agriculture Course?). Yet its development in the wine world is relatively recent. Let us take Nicolas Joly, for example, who has received much media attention since he began practising biodynamics in 1980 in his vineyard, La Coulée de Serrant, which is along the banks of the river Loire. In Alsace, I should mention Eugène Meyer and Pierre Frick. In Burgundy, a small group of pioneers including Jean-Claude Rateau in Beaune and Thierry Guyot in Saint-Romain have been growing vines biodynamically since the mid-1980s. After encountering them, some leading estates quickly altered their practices and began experimenting with the method. Domaine Leflaive, for example, began using biodynamics in 1990. At that time, however, biodynamics was not at all popular and these forerunners had to demonstrate great tenacity in implementing their convictions despite the disapproving eye and often hostile reactions from their neighbours.

Without overly exaggerating, let us say that at the very least these pioneers were seen as eccentrics. Even only a few years ago the image of 'organic' evoked a bearded ecologist, a hippie, with an image of non-conformist and perhaps even left-wing. This in itself was enough to frighten a generation of often conservative producers who had experienced the boom of modern chemical products over the last thirty years. These new products offered producers remarkably easy solutions. However, they remained unaware of the chemical drawbacks.

Today the situation has changed dramatically. First of all, the need for ecological awareness has become extremely important in all areas of society: food, industry, politics ... This need is also felt and shared by a new generation of mainly young winegrowers who do not have the same preconceived ideas as the previous generation. They are seeking ecological solutions that combine respect for nature with rigorous hard work. They have seen successful examples from the pioneers and hope to form their own opinion. For this reason many of them start experimenting with organic or biodynamic farming on part of their estate while telling themselves pragmatically that with time they will see which kind of viticulture suits them best. Statistics showing double-digit growth in total vine surface area cultivated organically confirm this trend in the most advanced regions.

For me, this trend is highly significant, and a step backward is unlikely in the coming years. It is important to realise, however, that with this trend comes a certain 'fad' effect which can sometimes be excessive. Some winegrowers claim to practise biodynamics but, from time to time, use a 'bit of' herbicide or penetrating fungicide ... For the consumer, there is only one answer to this: certification (see question 14: How does one know if a wine is biodynamic?). Although this is unfortunate, doesn't it represent in some way the 'price of fame', or at least a form of acknowledgment?

3 *Why is biodynamics spreading particularly in viticulture?*

It may seem trite to say, but wine is the one agricultural product that sparks the most interest, even passion, in regard to its consumption: from the qualities of its 'nose', its taste, the pleasure it can give, to the effects on the body and mind. Interest in wine has reached such a level that there are consumers who travel across the globe to see and touch the actual ground where the grapes are grown, and meet the artisan who grew and vinified them. This is indeed extraordinary given that the nineteenth century industrial model has progressively expanded to the rest of agriculture throughout the Western world! This trend has brought with it a range of adverse consequences including the large-scale standardisation of crops in order to reduce costs, the simultaneous neglect of soil's natural diversity, and unrestrained mechanisation which has left little place for a human role. This concept of standardisation and cost reduction originally developed at the heart of the automotive industry, but its irrational application in agriculture can lead to absurdities. In contrast, working on the land teaches us a range of skills and the importance of adapting oneself and being aware of soil's diversity in bringing out the best in each plot of land. In our post-industrial, overly-urbanised society, could wine be our last strong tie to the earth? As Colette so magnificently wrote:

> In the plant kingdom, the vine alone allows us to understand soil's true taste. How faithful it is in its translation! It senses the secrets of the soil and expresses them through the grapes. Flint makes known to us through the vine that it is alive and sustainable. The chalk cries golden tears that flow as wine ...

In tasting wine, everyone can feel and truly experience for themselves the impact that a certain form of viticulture has on quality and the expression of *terroir* (see question 13: Does biodynamics offer a better expression of *terroir?)*. It seems to me that if biodynamics is spreading at such a rate in viticulture, it is doing so to serve as a pioneer and driving force, in opening the door to consumer awareness of the drawbacks of industrial agriculture. And in a short space of time, the consumer will realise that what is true for the vine is true for all agricultural production.

Viticulture is all the more in need of organic and biodynamic farming as, along with arboriculture, it is the biggest consumer of phytosanitary treatments in France relative to surface area under cultivation. Vineyards in the Champagne region, for example, receive on average twenty doses of fungicides and insecticides per hectare, excluding herbicides (source: *Étude Agreste* 2006). This can be explained by the fact that the vine, like other fruit trees, is a perennial plant. As a result, it is more susceptible to parasites than annual plants which are rotated. And, to be honest, let us add that it is also the crop with the most added value which makes winegrowers a particularly attractive target for the phytosanitary industry.

4 *What is the difference between organic and biodynamic?*

Without going into regulatory detail, *organic* or *organic farming* is defined in practical terms as the discontinued use of products such as: herbicides, synthetic fertilisers, chemical insecticides, and/or fungicides that penetrate the vine's organs and sap (products labelled *penetrating* or *systemic*). Generally speaking, these are synthetic, petroleum-derived products. Their 'effectiveness' is unquestionable, yet so is their toxicity and side effects on the environment as well as human health.

To simplify, organic could be summarised as follows: *do not use the most toxic products.* It is important to note that organic farming (and biodynamic, incidentally) still allows for the use of certain products deemed less toxic such as sulphur and copper to treat vine disease (see question 8: Is copper toxic?).

During cellar tastings, when I meet wine lovers or even professionals who obviously have only a vague idea of what biodynamics is, I like to offer them a simple definition in three key points. First, biodynamics is organic and therefore begins by discontinuing the use of the chemical products listed above. In order to obtain biodynamic certification, one must first have organic certification. Yet, biodynamic viticulture is not limited to this; it requires a change in mentality on the part of the grower. This is a fundamental step which I will now explain.

Second, a biodynamic grower has a different view on disease than one practising conventional farming. This is similar to the different approaches to medicine between Western and various forms of traditional Chinese culture. Western medicine studies a disease in order to identify the pathogen (a bacterium, virus, fungus ...), then dissects the biology of the pathogen in order to develop a drug that blocks its development. From that moment on, once the disease appears, it suffices to administer the drug and thus suppress the disease ... until the next time. This is the logic behind the use of antibiotics, for example.

In biodynamics as in traditional medicine, however, a pathogen such as mildew is considered to be part of the natural environment. The pathogen appears because of an existing imbalance, and can develop at an overwhelming rate which may sometimes cause problems with respect to the quality and quantity of the harvest. Thus, a fungus is no longer seen as the cause of a disease, but as the result of something out of balance. In this case, it seems more important to focus directly on the cause of the imbalance, rather than on the symptom. Looking at disease from this perspective shows that 80% of imbalances come from diet, both in the case of the vine as well as humans. For this reason, biodynamics focuses on soil health (preparation 500 for soil structure and vitality, for example, and preparations 502–508 for the compost that maintains healthy soil). A vine can only be healthy and well-nourished if the soil is in a perfectly natural functioning state. Moreover, although the role of a soil's microbial life is still not fully understood, it should be treated with special care. In comparison, mineral fertilisers have the same effect on the vine as fizzy drinks do on people: unbalancing! All other preparations are often based on dynamised plant infusions, and are used by biodynamic growers to restore balance where imbalances are identified by attentive observation: lack of vigour, excess moisture ...

For me, the second point is essential in understanding the difference between organic and biodynamic. This difference is not really significant in its implementation, yet mentally, it may require a very profound rethinking of the old paradigm as well as considerable commitment and personal development on the part of the grower.

The third point focuses on observing and respecting rhythms, including the moon's rhythms, and will be developed at a later point (see question 9: What is the planting calendar and what is its purpose?).

5 *What about 'reasoned viticulture'?*

A few years ago, Western society quickly became decidedly aware of the need to change its behaviour and be more respectful of the environment and the Earth. As a result, several polluting industries put forward new, supposedly more environmentally-friendly approaches. In this regard, the consumer struggles to make sense of it and distinguish between a sincere, authentic change of behaviour and simple marketing, i.e. talk. 'Sustainable development', with all its trimmings is one example. Confusion reigns in the agricultural sector as well, and the term 'reasoned agriculture' (*agriculture raisonnée* in French) is largely responsible for it.

Introduced in France in 2004, so-called 'reasoned agriculture' is legally defined and its specifications are presented in the form of a list of 103 'requirements' (a Prévert-style inventory although less poetic, some might say) aimed at more respect for the environment. Yet, a closer look at the requirements shows that most of them simply reflect the basic regulations that have been in place for many years (e.g. using only chemical products approved in France), or respect for fundamental safety measures (wearing protective clothing when applying hazardous chemicals such as insecticides), or even simple commonsense when using pesticides (keeping up to date with journals that are likely to be filled with advertising for the same pesticides ...).

I personally believe in the sincerity and determination of a number of winegrowers engaged in this approach, and I must recognise the fact that it is a step in the right direction. In my opinion, the problem lies in the fact that it is a tiny step, and ridiculous in the face of the urgent need to correct agricultural practices that are so devastating for the Earth. What I fear is that producers as well as consumers will not be satisfied with such little progress. Instead of being a catalyst, reasoned agriculture will only be a hindrance to the changes needed in farming practices.

Let us be loud and clear: reasoned agriculture, which uses agrochemicals less irrationally and demonstrates slightly more awareness of their drawbacks, is nothing more than conventional agriculture.

6 *How does one fight vine diseases in biodynamics?*

In order to answer this question, I need to go back to a key point that I explained earlier in question 4 regarding the difference between organic and biodynamic. In biodynamics, disease is viewed differently. It is assumed that a pathogen (fungus, insect, bacteria ...) is only a symptom that appears because of an underlying cause or existing imbalance. Thus, the notion of 'fight a disease' must be replaced by a new logic of 'prevention and correction of imbalances'. Such reasoning suggests that if imbalances are reduced, the threat of disease should decrease. Vines are therefore better able to resist disease on their own, just like healthy individuals.

The primary source of imbalances in vines is precisely the accumulation of attacks from agrochemical treatments. What a contradiction! The products believed to 'protect' the vines are actually responsible in the long run for their increasing vulnerability to disease. And so from here the vicious circle of chemically-dependent agriculture begins: more chemical products lead to more disease which leads to an increase in the frequency and severity (and therefore, toxicity) of treatments, etc. It is often difficult to take the first step out of this logic because the winegrower is fearful of the risks that follow. One must break away from the use of chemical products. This is the most critical stage, known as the *conversion phase*, and lasts three years. During this time one must be extremely vigilant. Restoring balance is indeed a gradual process.

Let us go back to the logic which lies behind preventing imbalances. The great majority of preparations used in biodynamics are derived from plants. These plants are generally infused in water, then diluted, and finally dynamised (see question 12: What is dynamisation?) in a larger volume of water before being sprayed on the vines either with a hand-held sprinkler or a tractor. The logic behind these dynamised plant infusions could be compared with that of homeopathy: it involves a quantity of relatively diluted and dynamised plants. Unlike chemical products, these preparations do not act on a quantitative level, but rather on another level, let us say *energetic*. In any case, it is a level unrecognised by modern science.

It is important indeed to understand that the reasoning which guides the choice of preparations used is not scientific. It involves *analogical* or *symbolic* reasoning. Scientific logic analyses the causal chain, or the cause and effect relationship. Then, based on the principle that the same causes produce the same effects, one tries to understand and quantify the action mechanisms of various molecules. Conversely, analogical or symbolic reasoning leaves a great deal to intuition and the global perception of phenomena, without concern for the specifics of a mechanism (see question 10: Is biodynamics scientific?).

Let us take the example of horsetail infusion which is frequently used in biodynamics. Horsetail is a small plant that generally grows in very moist soils but which, surprisingly, has very restrained growth despite the abundance of water. Its structure is quite hollow and the stems and leaves look like hard, dry needles. They contain an enormous quantity of silica, rather than water. Horsetail is essentially the opposite of a water plant (the water lily for example, whose structure cannot even resist gravity). Thus, a brief observation shows that one of the essential qualities of horsetail is to successfully manage excess moisture, while remaining dry and maintaining its vigour. It is this quality of the plant that the winegrower seeks to transmit to the vine when he sprays the infusion. For this reason it is often used during wet springs. Horsetail shows the vine how to avoid excessive growth associated with an overabundance of water, which would otherwise lead to a significantly higher vulnerability to mildew. Intuition is not enough, of course. You need to test the procedure to confirm that the idea works. Such is the case with horsetail which is used by many growers.

The great majority of biodynamists continue to use copper and sulphur as a supplement in improving plant and soil equilibrium. Although these products are not produced synthetically and are less toxic and unbalancing, I consider them to still be part of the former 'fight a disease' logic. Copper is principally used to prevent downy mildew while sulphur is used against powdery mildew (here I am referring to sulphur sprayed on the vine, and different from sulphites used in the winery during vinification and which will be discussed in question 17: Does a biodynamic wine contain sulphur?). Even if most biodynamic growers set increasingly strict targets to reduce the quantity of copper and sulphur they apply, their use is still based on quantitative logic, which is to say, adapting the dose to the severity of the disease.

It should be noted that certain producers, in certain vintages, on certain parcels, succeed in growing their vines without copper or sulphur. In my opinion, these experiences are and will be the main subject of viticultural research today and for years to come.

7 *Is pollution from neighbouring vineyards a problem?*

This is probably the most frequently asked question. And, when you see how vineyards are parcellated in Burgundy or other regions such as Alsace, it is indeed a justified one. Domaine Zind-Humbrecht in Alsace, for example, farms approximately 40 hectares spread out over some 100 parcels. This is a far cry from the winegrowing estates in the Médoc with several tens of hectares in a single plot. Is it not a bit far-fetched, therefore, to think that it is possible to grow grapes without chemical products, when the owners of neighbouring properties do not farm organically and spray synthetic pesticides on their vines and sprinkle herbicides on the soil? Moreover, I must remind you that the width between two rows of vines in Burgundy is only one metre ...

Well, in practice it is not a problem!

First, there is very little product drift from tractors equipped with modern spraying methods. Herbicides stay on the targeted soil, but then leak into the surface or penetrate the water table. Products sprayed on leaves slightly contaminate the row that borders the neighbouring property, but rarely much further. Obviously this is not the case when proper spraying practices are ignored: spraying in very windy conditions or by helicopter. Fortunately such cases are the minority.

Secondly, whatever the product used, there is always a small percentage that vaporises into the atmosphere or leaks into rivers. Such residual pollution can spread quite far, from several hundred metres to several tens of kilometres. In the world we live in, both the water and the air are often polluted. This is particularly true in dense, industrialised areas. Thus, it would be absurd to expect to grow grapes without any exposure to pollution, unless you live under a glass dome.

Thirdly, vines grown biodynamically are even less sensitive to this small amount of residual pollution as they have not been destabilised by regular exposure to large quantities of chemical products. As I explained in question 4 (What is the difference between organic and biodynamic?), biodynamics goes beyond simply renouncing polluting products. Working in harmony with life forces, it aims to grow plants with optimal equilibrium. These plants are thus capable of resisting such small disruptions themselves. Perhaps even the rows belonging to the neighbour also benefit from the positive effect of biodynamic preparations ... And when I say that in practice the form of viticulture carried out by neighbouring growers is rarely a problem, I would emphasise the fact that this difference in form between two producers is generally visible in the very first row. It can be observed by the colour of the leaves, the general appearance of the plant, and the condition of the soil.

However, to be completely honest, I must admit that I sense that the form of viticulture practised in neighbouring vineyards is not entirely without impact on the health of those parcels grown organically or biodynamically. To this day, conventional agriculture, with all of its consequences in terms of pollution, imbalances and disease, remains the most practised method of production. As a result, the environment is out of balance on a large scale. And, generally speaking, plants get sick when their environment is disease prone. I think it is for this reason that many of the estates practising biodynamics continue to be confronted with the threat of disease, and therefore still need to use copper and sulphur. In relatively advanced regions such as the *Côte de Nuits* or the *Côte de Beaune,* I am very hopeful that within the next ten to twenty years, organic and biodynamic viticulture will be practised by the majority. Then, residual pollution will decrease significantly, and growers will be able to forgo the use of copper and sulphur completely.

8 *Is copper toxic?*

Critics often reproach organic farmers for the following: by having to refrain from using synthetic chemicals, they instead have to use large quantities of copper to fight disease, especially mildew. And, since copper is a heavy metal, it would accumulate in the soil and its toxicity would be responsible for environmental pollution that would be even worse than that caused by synthetic chemicals.

This remark must be taken seriously as copper is toxic in high doses. A few decades ago, it was not unusual to see growers spraying annual doses of Bordeaux mixture (*bouillie bordelaise*) equal to more than 10 kg of copper per hectare (ha). Today, organic specifications limit copper use to an average of 6 kg/ha per year over five years. Some biodynamic producers associations are more strict: Demeter aims for 3 kg/ha per year on average over five years. These growers are aware of copper's drawbacks and are constantly seeking ways to minimise its use. But for me, when copper is used in moderation, there is no comparison between it and synthetic pesticides in terms of harmful impact.

First of all, copper is a *contact* product. In other words, it stays on the surface, on the outside of the leaves and the grape. It forms a sort of barrier that stops attacks from the fungus responsible for mildew. Chemical fungicides, however, are generally *penetrating* or *systemic*. Penetrating products, as their name suggests, penetrate plant cells and act at the biochemical level. Systemic products go even deeper into the plant's system to the sap. From there, they spread to all the organs, to the grapes and even the roots. Clearly their effectiveness is far superior to that of copper ... as is their potential for disrupting the vine's balance. In addition, being a heavy metal, copper is eliminated after pressing during clarification. On the other hand, traces of synthetic products can be found at the biochemical level of grape cells, and therefore in the wine. For the consumer, the difference is indeed substantial.

Secondly, it is important to add that copper is generally used in the form of a mineral salt, for example copper sulphate. In very small doses it is a trace element and, contrary to chemical products, essential for life. In humans and mammals for example, who are regulated by the liver, copper plays a role in the functioning of the immune system. Copper is stored, excreted in the bile and distributed to the organs. We are talking about doses of a few mg/kg. Its anti-infective properties were already known and used in ancient Egypt and are still used for their purifying properties in certain cases: roofs or gutters to collect rainwater, piping, and preserving pans for jam. Thus, even in moderate doses, it is toxic for microorganisms such as mildew, and it disrupts soil's microbial life. It is all a question of dosage.

Let us make a quick calculation: one hectare (ha) of vines, 10,000 m2, receives an average dose of 3 kg of copper annually. Let us suppose that this copper is deposited entirely in the soil which has an average depth of 50 cm and a volume of 5,000 m3/ha. Soil density is approximately 1,200 kg/m3 (considerably higher for land in Burgundy which is rich in clay and limestone), thus approximately

6,000 tonnes/hectare. According to this hypothesis, the use of copper would be equal to 0.5 mg/kg, which is a concentration relatively compatible to humans. Reputed soil microbiology specialist, Claude Bourguignon, estimates that, depending on the nature of the *terroir,* a well-functioning soil can eliminate approximately 1 to 3 kg copper per hectare per year via metabolisation.

For this dilemma as for all, it is a question of moderation. From my perspective, I believe that for many biodynamic producers, the amount of copper used is compatible with the maintenance of a living soil (which the land itself displays), and is a thousand times more desirable than the use of synthetic chemicals.

9 *What is the planting calendar and what purpose does it serve?*

The sowing and planting calendar is the tool used by biodynamic growers to work in accordance with the *rhythms.* Published annually in English by Floris Books, it is the fruit of Maria (who died in 2012) and Matthias Thun's work on lunar and planetary influences on sowing and planting.

One of the important aspects of biodynamics is that a winegrower does not simply farm his own small, isolated parcel of land. He is part of a far greater ecosystem from which he can perceive influences that are of interest to him, and eventually take them into account in his work. He is part of the local *terroir,* the greater region, the planet Earth, the solar system and, more generally, what the ancients called the cosmos.

At first, the notion that planets can influence us is generally difficult to accept, especially in Western culture. With the exception of a few cases such as the moon's influence on tides, modern science does not explain, and therefore accords no value to many of the influences described in various traditions. Yet, even in our society, many people are challenged by this approach, and it is often what people remember about biodynamics: '*Oh right, you plant with the moon!*' For those who cannot imagine such planetary influences, as if they were able to send us a ray of something, I would offer the following hypothesis. Let us assume that a larger environment exists which covers us all, including the planets whose rhythms and cycles influence us. The planets serve only as a guide, as points of reference in the cycles, often like the hands of a clock. Knowing the moment of the cycle enables us to predict the likely influence.

Once again, in order to understand these cycles, biodynamics requires symbolic reasoning. For example, everyone distinctly perceives the cycles associated with the sun: the rhythm of the hours of the day and seasons of the year. The duration of the months originates in the moon's rhythms which can be compared to those of the sun. An ascending moon is a *mini-spring*. The moon travels higher and higher in the sky, just as from Christmas to June 21 (from the winter to the summer solstice) the sun at noon gets higher and higher in the sky. As a consequence, one can say that this favours the forces of expansion or externalisation. On the other hand, a descending moon represents a *mini-autumn* which favours the forces of internalisation, or the process of storing energy. Biodynamics often uses the processes of expansion/contraction, outward/inward, growth/limitation, inhalation/exhalation. Grass, for example, cut during an ascending moon could tend to grow more, and thus come back more quickly. Yet, when cut during a descending moon, it will grow back less quickly.

There is another lunar influence mentioned and used considerably in the sowing and planting calendar: *Fruit, Root, Flower* and *Leaf* days. In essence, these influences are associated with the position of the moon with respect to the constellations of the zodiac. They represent 1 to 3 day periods which are particularly favourable to the development of one of the four parts of the plant. The most important agricultural operations (sowing, ploughing, spreading biodynamic preparations, harvesting) should be carried out based on which part of the plant is destined for consumption: a Leaf day for salads, for example, a Root day for radishes, and of course, a Fruit day for vines grown for their grapes.

The influences of these rhythms are important and observable, provided you pay attention to them. It is necessary, therefore, to be aware of them and use them. However, this does not mean forgetting the rest, that is, good agronomy and farming practices. It would be absurd to plough on a Fruit day if the soil were too wet or too dry. The result would be disastrous. It would also be useless to harvest in the rain even if it were a very favourable Fruit day. For me, it is important to take the elements of the sowing and planting calendar into consideration as they allow for more discerning work, more in harmony with nature. Yet, you should not place undue importance on the calendar, and it must be one deciding factor among many. Everyone needs to set their own priorities, and this is what constitutes the know-how and talent of a grower!

Various planting calendars are published and all are based on the same astronomical data. The specificity of each, however, lies in the interpretation and ranking of the astral influences they propose. As well as the planting calendar by Maria Thun, many biodynamic associations in various countries have their own calendar.

10 *Is biodynamics scientific?*

When talking with some scientists, professors or researchers about biodynamics, there is often a quasi-epidemic reaction of more or less aggressive rejection. Having said that, certain biodynamists sometimes react with at least as much hostility to scientistic discourse.

I personally became aware of this division a few years ago. I had just arrived at the Montpellier School of Agriculture where, as a simple wine enthusiast, I was going to resume my studies in oenology. On the first day of class, the distinguished professors, who were going to instill in us their knowledge over the course of the coming months, had gathered together to present the programme. At the end of his presentation, the Professor of Viticulture concluded in a joking manner: '*... and of course, within these walls we will not talk about biodyn- ...' Oh no! No witchcraft here.'* A burst of laughter followed. The message was clear. Mindful of the Cartesian principle that to free yourself from prejudice, you must employ scepticism, doubt everything and be open to every alternative, as you can only accept something as true if it is based on your own experience, I promised myself that I would quickly look into this other form of viticulture.

With my knowledge of both the scientific world and biodynamics, I would like to offer a few thoughts to reflect on in order to reconcile the two camps.

Let us say straightaway, *biodynamics is not scientific*. In any case it is not scientific in the sense of applying scientific reasoning which only recognises truth that can be proven by scientific logic (such as mechanics or biochemistry). Scientific logic generally relies on precise, objective and quantifiable measurements, and reveals a chain of causal links. Biodynamics, on the other hand, leaves much to intuition, analogical and symbolic reasoning, which is normally used by artists and poets. One example of symbolic reasoning (if rather a caricature, I admit) would be: 'This flower is yellow, therefore it makes me think of the sun, and the sun warms, therefore when the sun is not shining, this plant can offer me feelings of warmth.'

It is important to understand that these two approaches involve completely different ways of reasoning, each with its own logic. It seems just as absurd to want, at any price, to apply the logic of one to judge the validity of the other: thinking that one is more valuable or superior to the other. In Western society, scientific reasoning is highly developed and has brought us all the technical progress we know. Consequently, there is hardly any truth apart from Science! This represents the same distinction which is often made between the two hemispheres of the brain: analysis, quantification, and organisation on the left brain; intuition, sensitivity, and imagination on the right brain. Yet, we would all gain by simultaneously developing these two poles of our thinking. Society would then find itself more in balance.

Biodynamics focuses on the life of plants and animals. In this realm, analogical reasoning can be powerful and fast. Biodynamics assumes that the laws of physics alone do not explain all the subtleties of living things. Other forces exist which are not directly accessible by our five senses, and are, therefore, immeasurable. They are nonetheless important in agricultural work. Rudolf Steiner wrote:

> A view based solely on facts supplied by our physical senses is not far-reaching enough ... spiritual research is more concerned with spiritual causes lying behind matter ...
>
> As an example, if there is a person in front of us who is raising a hand, this can suggest two different ways of looking at this action. We can either investigate the mechanism of the person's arm and the rest of his or her body and describe the process that is taking place as a purely physical one, or we can turn our spiritual attention to what is going on in the person's soul and to his or her inner motivation for raising a hand. (*Esoteric Science,* p. 119)

In other words, if modern science focuses on the study of physical phenomena and attempts to describe each perfectly, it is perhaps more limited where the understanding of life forces is concerned. In a recent interview on the radio, I heard Serge Haroche (French Nobel prizewinner in Physics in 2012), lay great stress on this limit of modern science. It is precisely this additional dimension that biodynamics seeks to formalise and introduce in agriculture.

11 *In what way is biodynamics more respectful of nature?*

This question stems from the response to the preceding question. Conventional agronomy does not understand the subtle laws that govern Life, or Nature, and is only interested in that which concerns the most material of physical aspects: mechanics, chemistry, and so on. The word *agronomy* stems from *agros* (field) and *nomos* (law). Does it mean that the agronomist wants to understand the laws of the soil, or to impose his law on the soil?

It is for this reason that the chemical struggle is destined for failure. The solutions it proposes are often *against nature*. There is no other possible long-term outcome than a worsening of the situation. On the other hand, the etymology of biodynamics expresses interest in the subtle forces behind matter. In German, biodynamic was originally *biological and dynamic agriculture*. Biological: no need to explain. Dynamic: *dynamis*, force. Thus, in addition to biological agriculture which refuses products opposed to life, biodynamics studies and uses the forces and laws of the living. Unfortunately these forces are not measurable, but their effects can be observed. Optimising these forces is the aim of all biodynamic preparations as well as dynamised infusions.

12 *What is dynamisation?*

This is a delicate question that goes to the heart of biodynamics. Explaining dynamisation is a difficult task for me given that I feel the need to deepen my own knowledge of the subject, with aspects still to be explored and tested. Nonetheless I will try to convey to you my current understanding of dynamisation.

One of the principles of biodynamics assumes that the surrounding environment is made up not only of matter, but also of structuring forces and principles not directly perceptible to our five senses. Accordingly, it is more efficient to work with matter (chemistry and conventional mechanics) and these other forces and principles simultaneously, especially when dealing with living organisms as in agriculture. Dynamisation is a key phase in working with these 'subtle' elements. It is important to note that the common etymology of biodynamic and dynamisation, as we saw above, comes from the Greek *dynamis,* or force.

When referring to this principle, some people use the term 'energy'. In biodynamics one talks about 'ethereal' or 'astral' aspects. To simplify matters, I will use the generic term 'information' here to indicate that a material substance is a transmitter of information related to a particular force or process.

Functions of dynamisation

Outside biodynamic agriculture, dynamisation is used particularly in the preparation of homeopathic medicine at each step of dilution.

First and foremost, it involves the vigorous mixing of a generally small amount of an 'informed' substance in a volume of water. At first glance, this represents the blending and homogenisation of a diluted solution. But beyond this simple quantitative effect, it entails, above all, the transmission of information and the infusion of energy at several levels.

At the first level, it involves freeing the information carried in the diluted substance and transmitting it to the entire volume of water in order to multiply its effect and enable distribution over a large surface. For example, a few grams of horn silica preparation #501 (2–4 g) are used in several tens of litres of water (30–100 lit.) to treat one hectare of crop. At the end of dynamisation 'informed water' is obtained – that is, 'informed' by preparation 501 – and this should be sprayed within two to three hours. Clearly, high quality water is essential for good dynamisation. The grower will thus opt for water free of dissolved pollutants (chlorine, residual pesticides, and so on) but also 'living' water that is highly receptive to the information. For this reason a large number of winegrowers use rainwater for dynamisation.

At the second, more subtle level, the water is imbued during dynamisation with information from the global environment. It involves physical elements that are directly in contact with the water and the immediate environment which must be carefully selected: copper or stone dynamisation tanks, absence of electromagnetic pollution ... In biodynamics, the environment also refers to the current cosmic configuration (the position of the sun, the moon and the planets with respect to the Earth). The dynamisation has a specific duration, usually one hour, and the cosmic configuration will also permeate the mixture during the mixing process. For this reason, the winegrower will carefully choose the timing of the dynamisation (the beginning or end of the day, descending or ascending moon, Fruit day ...).

At the third level, human thought is absorbed and influences the water. This point touches on the vital importance of thought and its effect on matter and life. The winegrower should therefore carry out the dynamisation very attentively. In other words, he should be completely aware of what he is doing and the reasons for his actions and objectives. It should always be done with full attention and concentration. I invite you to read the interview with Anne-Claude Leflaive by Sylvie Ogereau, which appeared in the February 2010 edition of the *Revue du Vin de France*. In it she explains the importance of *enlivening* harvesters so that they will be in a good mood and ultimately transmit it to the grapes.

The impact of man's attitude and thoughts certainly has greater influence than modern Western society accords it. I have noticed that Eastern society is culturally more sensitive to this idea.

Types of dynamisation

Dynamisation carried out in biodynamic agriculture involves mixing water in a vigorous circular motion to create a whirling effect, or a vortex. Once the vortex forms, it is mixed in the reverse direction resulting in a state of chaos and then followed by a vortex in the opposite direction. This alternating rhythm vortex/chaos, which symbolises the densification and structuration of the matter, followed by its dissolution and reorganisation, is linked to the ancient alchemy principles of *coagula* and *solve*.

Dynamisation can be performed by hand (with the arm or by using a stick) for small quantities, or by machine for larger volumes. Winegrowers frequently use dynamisers with 150 to 300 litre tanks, and the dynamisation process generally lasts for one hour. In the beginning, Rudolf Steiner is said to have carried out the first dynamisation demonstration using a bucket and his cane.

There are a number of other dynamisation methods including the following. First, there is dynamisation by succussion, which is done by shaking the container and is used in homeopathic laboratories. There is also dynamisation by circulation of liquids in conduits or basins which produces currents or a special whirling mass (flow forms, for example).

13 *Does biodynamics express* terroir *better?*

Let us return for a moment to the concept of *terroir* which is so often used in viticulture, yet always debated. The French term *terroir* does not really exist in other languages, and particularly not in English. For thirty years even prominent university professors have been trying more or less successfully to scientifically conceptualise this term. Essentially, it is a term with a very rich meaning, but particularly vague around the edges. For this reason, everyone understands it, but no one explains it ...

The term *terroir* essentially expresses a place, a location in nature with a specific climate, a soil on which a plant grows, a taste, the 'taste of the soil' as Colette would say, as well as an artisan, the one who grows the plant on the soil, who observes, tastes and feels it.

Then, of course, the definition that modern agronomy proposes: the useful understanding of physical (granulometry, porosity ...) and chemical (nutrients, trace elements ...) compositions of the soil. You should remember, however, that in Burgundy for example, with our sophisticated, modern methods of analysis, we do not do any better than the ancients or the monks who, over the centuries, so rightly defined the Burgundian *climats*. Today, the only thing we can say is: 'They were right ...' How did they do it? Can one talk about the 'energy of a place', or the importance of the underground bedrock as geobiology does? Orientals often talk about the Earth's *chi.*

In any case, it seems to me that the concept of *terroir* suggests a particular idea of the right gesture made by the winegrower who is in harmony with the place, the plant and nature. From this perspective, it appears that chemical agriculture, which denies this search for *rightness* at the heart of nature, is largely unaware of this dimension. Chemical agriculture is therefore *wrong* from the point of view of *terroir* expression. On the other hand, biodynamics, with this added dimension of respect in understanding nature, is currently the most advanced form of viticulture in the search for rightness.

It is difficult to know exactly how *terroir* translates into taste, or a particular sensation in the wine. Yet, the following elements constitute at least some key points:

- Plant nourishment through its roots requires the best possible functioning of the soil. Respect for its vitality and its microbiological richness alone guarantees the continued existence of its physical and chemical properties. This seems evident. Unfortunately, chemical agronomy has largely forgotten it.
- Optimum plant health and general equilibrium are sought. Limiting potential imbalances is achieved once chemical pesticides are abandoned and replaced by biodynamic preparations and plant infusions.
- A grape's natural balance at harvest (sugar, acidity, aromas, and tannins) renders useless the array of modern oenological techniques aimed at correcting the wine during vinification. Any alteration necessarily reduces faithfulness to *terroir.*
- Populations of indigenous yeasts naturally present in

> each vineyard parcel can play their essential role during fermentations when the taste of *terroir* reveals itself. From this point of view, the processes at play during fermentations remain largely a mystery and are not limited to the biochemical transformation of sugar into alcohol. It should simply be noted that the differences between two *terroirs* are far more evident in the final wine than in the grape.

It seems to me that, based on at least these four points, biodynamics is the form of viticulture that facilitates the most faithful expression of *terroir.*

Part 2

VINIFICATION

14 *How does one know if a wine is biodynamic?*

Since approximately the mid-2000s, biodynamics has been fashionable in the wine world. Today, a very interesting phenomenon can be observed: whereas fifteen years ago, biodynamic growers were seen as eccentrics, today biodynamics has become a marketing tool for some. The ensuing risk, however, is talk which sometimes surpasses what is really practised in the vineyard. Numerous advertisements make reference to respect for the Earth, the moon, the cosmos ... How can a wine lover really know what form of viticulture the grower actually practises?

The goal of certification is: 'Say what you do, and have verified that what you do is what you said you would do'. Concretely, biodynamic producers associations establish standards (Demeter or Biodyvin). Growers who wish to obtain biodynamic certification become members of the association and agree to comply with the standards. Then, each year, the association commissions an independent inspection body (Ecocert, for example) to go and verify that the grower is indeed in compliance with the standards. Thereafter, the association annually issues a certificate of biodynamic viticulture that the grower can show to clients. The producer can also display the association's name on their product label. It is essentially the same procedure one follows to obtain the organic standard label AB *(Agriculture*

Biologique). However, the standards for organically-produced products are defined by central government and the European Union.

Biodynamics is, of course, an agricultural technique and therefore applies only to grape growing. It is for this reason that, along with organic wines, the following type of indication has been found on the label up to the 2011 vintage: 'Wine produced from biodynamically grown grapes'. Yet, producer associations have recently started asking their members to more closely observe the minimum requirements regarding vinification. These rules aim to extend respect for the natural raw material to the winery, and eliminate the most 'aggressive' oenological practices (see question 16: in the winery: does biodynamics apply to vinification?). In addition, new European organic wine standards which cover the rules concerning vinification, have been adopted on February 8, 2012, by the European Commission and take effect at the beginning of the 2012 vintage in France. Until now, only grapes were certified as 'organically grown'. The new standards allow growers to substitute this with 'organic wine' on their labels.

In reality, a significant number of growers who cite biodynamics as the form of viticulture they practise, are not certified for various reasons. Certain long-term practitioners are absolutely resistant to external inspection and to every additional administrative procedure. Some grow only part of their vines biodynamically. Still others simply take advantage of its commercial trend, and the pretext of wanting to maintain one's freedom sometimes serves as a cover-up for viticultural practices unauthorised by the standards.

For the wine lover who cannot personally verify a grower's practices, certification remains the only guarantee.

15 *What is the difference between the various biodynamic producers associations?*

Over the last few years, biodynamics' rapid development at the heart of viticulture has led to the creation of several producer groups. In France, three main bodies currently represent biodynamics: Demeter, Biodyvin and Renaissance des Appellations, or Return to *Terroir.* Naturally, the number of associations alone carries with it a certain risk of confusion.

It is important to know that since its inception, the biodynamic movement has been very structured and centred around the official bodies headquartered at the Goetheanum in Switzerland. The Demeter brand, in particular, has been used since 1928 to create brand awareness among consumers. Today there is one global association, the International Biodynamic Association (IBDA) which unites all biodynamic agricultural activities in each country. Additionally, Demeter-International oversees a network of national Demeter organisations in sixteen countries, including Demeter-France, Demeter Association Inc. (USA), and Demeter UK. Their primary task is the compilation of standards for biodynamic producers, as well as the organisation of inspections for certification and commercial use of the Demeter brand on products. In concrete terms, Demeter is the official certification body and guarantees consumers producer conformity to accepted biodynamic practices for all agricultural crops including grapes. For practical purposes, Demeter delegates inspections and the organisation thereof, to an independent, specialised firm (Ecocert, for example) who simultaneously inspects the organic agriculture standards application. The Demeter-France head office is located at the *Maison de la Biodynamie* in Colmar, Alsace.

The *Syndicat International des Vignerons en Culture Biodynamique* (SIVCBD, known as *Biodyvin*) is another producers' association created in 1996, and is the result of a split at the heart of the biodynamic movement. At the time, certain biodynamic purists regularly accused winegrowers of not fully complying with all biodynamic principles (notably the one regarding the farm as an *agricultural organism*), citing, in particular, single-crop farms and the absence of animals. Given the fact that at the time membership fees were based on producer revenues, and winegrowers were by far the biggest contributors, it is easy to understand why some growers did not appreciate the recurrent criticism. Thus, a small group of *domaines,* led by the Alsacian Marc Kreydenweiss and the consultant François Bouchet, decided to create their own association focusing uniquely on viticulture, with its own standards and its own certification under the *Biodyvin* brand. Biodyvin is exclusively comprised of biodynamic wine producers who share the same philosophy: the pursuit of excellence in wine quality as well as the respect for the expression of *terroir.* New members are only accepted if they share this commitment to quality. For this reason, Biodyvin is also a place for technical exchanges among élite growers, and a consumer guarantee uniting *biodynamics* and *great terroir wine.*

Like Demeter, Biodyvin commissions the actual inspection process to the independent firm, Ecocert.

Over the last few years, Demeter and Biodyvin have each been working on the establishment of standards which also cover winemaking operations (see question 16: Does biodynamics apply to vinification?).

The third group, *Renaissance des Appellations,* or Return to *Terroir,* was created more recently in 2001 by the well-known biodynamic pioneer from Anjou, Nicolas Joly, famous for his wit and his writings (*Wine from Sky to Earth,* published by Acres USA). The aim of this association is the promotion of biodynamics, and, more generally, a specific philosophy of wine based on the pursuit of the strictest authenticity in the vineyard and the winery. In order to have the greatest market impact, promotional activities are essentially carried out in the form of trade fairs for professionals (importers, sommeliers, wine merchants, and so on) and wine lovers from around the world. *Return to Terroir* is rather a communication body for biodynamics and is not involved in the inspection or certification of its members who should first be certified by Demeter or Biodyvin.

Today, with the increasing success of biodynamic viticulture, there is the risk of this term being abused. Marketing and communication can sometimes get ahead of agricultural reality. I remember seeing in 2009 a full-page advertisement in the *Revue du Vin de France* of a producer using biodynamic viticulture as a sales tool, even though he was not certified. For wine enthusiasts, it is difficult to really know what happens in the vineyard, and to distinguish between authentic biodynamics and counterfeits. Contrary to organic farming, biodynamics is neither recognised nor defined by central government or the European Union. Therefore, the only possible assurance today remains recourse to a trademark. For this reason, the terms 'Bio-Dynamic' and 'Biodynamic' are trademarked by Demeter. However, so far the association has never sued an unscrupulous producer for trademark infringement in order to protect its name. This is not how Demeter operates. It relies on consumers to be aware of what is actually practised in the vineyard. This is a word of caution, though, to wine lovers and journalists alike.

16 *In the winery: does biodynamics apply to vinification?*

Biodynamics in itself is a form of agriculture and therefore pertains to what happens in nature, or in this case in the vineyard. As with organic farming (in France labelled AB), its certification standards initially applied only to grape growing (see question 14: How does one know if a wine is biodynamic?). However, since 2009, the two certification bodies Demeter and Biodyvin have asked their members to adhere also to vinification standards that include wine-making tasks from harvesting to bottling. What can these standards consist of, given that neither Rudolf Steiner nor his successors in the anthroposophy movement provided systematic instructions concerning the transformation of food, in general, or wine, in particular?

Vinification standards seek to guarantee customers the preservation and enrichment of the qualitative value of a food, and more generally, respect for all the qualities of biodynamic agriculture over the course of the winemaking process. In theory, the standards aim for the following ideal: 'a wine made from biodynamic grapes, with nothing added during winemaking, *élevage* or bottling'. In practice, as with organic wines, the standards only restrict the use of modern oenological techniques and additives: the limited addition of sulphur (a little ...), the prohibition of industrial yeasts, the prohibition of acidification, chaptalisation limits, and guidelines for fining and filtration, to cite just a few. In short, this very much reminds me of the 'negative' definition of organic: prohibit the most aggressive, unbalancing and even harmful chemical and physical practices (see question 4: What is the difference between organic and biodynamic?). Obviously you would not want to destroy in the winery the work done in the vineyard.

Nevertheless, I would like to make three comments as I find these standards rather incomplete.

First, the debate concerning the authorisation or non-authorisation of certain practices continues to upset growers rather intensely. Let us take chaptalisation, for example (the adding of sugar to unfermented grapes), which some consider unnatural as its purpose is to alter the initial balance of the grapes and, as a result, it aims to standardise vintage characteristics, which would ultimately dismiss all respect for *terroir.* Others, however, believe that the moderate addition of organic cane sugar during fermentation entails an increase in the wine's alcohol content, which would be positive. The alcohol would help to develop the natural forces present in the grape, and thereby even significantly improve a wine's ageing potential, while being harmoniously integrated into its structure. This debate has not been resolved yet.

Second, exemptions are allowed for most of the outlined practices (acidification, chaptalisation, the addition of yeasts, and the addition of sulphur ...). How can one maintain credibility under these conditions? Credibility is all the more tarnished as you can find examples such as Pierre Overnoy or Marcel Lapierre who make an 'ideal wine' (without any additives) generally without even referring to any standards.

Third, the current standards constitute a 'negative'

definition of winemaking. In other words, it is a list of prohibited practices, which, in my opinion, can only be a first step, waiting for a second. A second step would involve the development of true biodynamic vinification operations including the observance of cosmic rhythms (Biodyvin briefly mentions this in its standards), the development of dynamisations for use during vinification, informed waters, and bioenergetic techniques. A small number of pioneers have already been experimenting with this for several years. Hopefully this know-how will spread sooner rather than later.

Keep in mind the anthroposophic conception of food which is to nourish the physical body, and also the life of the soul and the mind. Thus, it is natural that biodynamics must also concern itself with the transformation of harvested fruits. Nevertheless, this aspect is still relatively unexplored and we should be careful not to excessively or prematurely constrict individual initiatives.

17 *The use of sulphur: does a biodynamic wine contain sulphur?*

First of all, in order to avoid any confusion I would like to briefly clarify the difference between sulphur applied in the vineyard and sulphur used in the winery.

The first is applied as an anticryptogamic (fungicide) element to fight powdery mildew in particular, and exists either in a spray form (powdered sulphur), or in an even finer, micronised form (sulphur soluble in a liquid spray). This sulphur adheres to the leaves of the vine for a few days, and, during the process of sublimation (transformation from a solid to a gaseous phase), it releases vapours that interfere with the metabolism of the fungus.

The second is used to stabilise wines. In this case, the term *sulphites* is employed. This is obtained by burning elemental sulphur which releases the gas sulphur dioxide (SO_2), and is similar to the burning of carbon molecules which release carbon dioxide (CO_2). As it is quite irritant and impractical to use directly, it is generally dissolved in water to obtain sulphites (H_2SO_3). It is this watery solution that the winemaker can add to the wine during vinification and bottling. It is of course only the second use that I will discuss here.

What are the properties that make sulphur dioxide so indispensable? It is a remarkable preservative that protects the wine against two major risks which threaten the wine over the course of its life: microbiological contamination and oxidation. Thus, SO_2 is both an antiseptic (antibacterial and antifungal depending on the dose) and an antioxidant. Yet, used in large doses, it leads to serious problems:

- For the wine: it can completely block aromatic expression and harden the palate. It is then difficult to talk about expression of *terroir* because sulphur smells like sulphur whether you are in Sancerre or New Zealand ...
- For the consumer: Be careful! A headache is guaranteed!

In biodynamics, vinification standards seek to prevent the excessive use of additives, sulphites in particular, which could reduce the original energies present in the grapes. The objective is to administer a dose which corresponds to the actual need of the wine. This is truly the work of an artisan, determined on a case by case basis: wine by wine, vintage by vintage. For this reason, it is absolutely impossible to formulate a general rule for all, and standards simply set a maximum limit which is more restrictive than the law. Yet, you will find substantial differences in the quantities of added sulphites. If you compare two biodynamic producers, some even succeed in regularly doing without it altogether. The following list is an indication of total sulphur dioxide limits according to various 2011 standards. I have taken the example of a white Burgundy: a dry, white wine with long barrel maturation.

- General regulation (European): < 200 mg/lit.
- European regulation, organic wine: < 150 mg/lit.
- Demeter: < 90 mg/lit., exemption possible: < 140 mg/lit.
- Biodyvin: < 135 mg/lit.
- *Association des Vins Naturels* (AVN): < 40 mg/lit.

The *Association des Vins Naturels* is a body of producers and amateurs who share the common objective of winemaking 'without sulphur', and imposes a very strict limit. For the most part they are in favour of organic and biodynamic viticulture. However, as there is no self-imposed inspection or certification, it is up to you to trust them or not.

By now, you will have understood that *biodynamic wine* is not synonymous with *wine without sulphur*. Although some pioneers are showing the way, they sometimes do so at the price of high risk. Numerous excellent biodynamic growers (in particular producers of the finest whites) realise that, at this point, if they were to eliminate sulphur completely, they would not be able to consistently produce such delicate, mineral wines true to their *terroir*.

18 *Does a biodynamic wine age well?*

I have just addressed the question of sulphur, directly linked to that of conservation. At this point the following should be clear: for most winegrowers, practising biodynamics as well as seeking to decrease sulphite doses, corresponds to the pursuit of the highest authenticity, the greatest faithfulness to *terroir,* and the highest respect for the natural energies of the grape and the wine. We can thus expect a higher level of vitality and energy when the wine is bottled.

Biodynamically produced wine will therefore be alive, and normally more alive than a wine made conventionally. Consequently, it will also be more sensitive to surrounding storage conditions, favourable or not. First of all, there is temperature. All wine lovers know this, but in practice, very few really take it seriously. Let us remember that a biodynamic wine without added sulphites must be stored at a constant temperature of 15°C or lower. Ideally, for a cellar to guarantee optimal ageing it must have a temperature of 12°C +/- 2°C. Cellars of such quality are rare. Thus, for the most part, storage should be temporary, in other words, less than two years.

All other disturbances are potential sources of fatigue for the wine:

- Noise and vibration (often in a city);
- Chemical pollution (foul smells or chemical product residues in the place of storage);
- Electromagnetic pollution (electric power devices or high-frequencies);
- Geobiological disturbances

Irrespective of these considerations regarding the place of storage, I must also say a word about bottle closures. Cork is the traditional form of closure. As it is a natural material, its properties are inevitably inconsistent. There is indeed a considerable variation in wood fibres within the cork: density, age, the presence of scabs or holes, ripe or green tannins, not to mention residual chlorine pollution which is responsible for cork taste. Even in strictly identical storage conditions, variations exist from one bottle to another, unavoidably intensifying the longer the bottle is kept. The well-known saying among sommeliers comes from this: '*There is no great wine, there are only great bottles!*' These observations have led some biodynamic growers who vinify without sulphites to opt for other closure techniques quite some time ago: metal capsules, as used by Pierre Frick in Alsace, or even glass stoppers. At first glance this can seem shocking because it is less natural. Yet, time will tell if the wine's energy is indeed preserved over the long term.

For all these reasons, some might consider biodynamic wine to be more temperamental. As it is more sensitive to its surroundings, a biodynamic wine will probably be more unpredictable than a wine made by chemical agriculture which would be more rigid. Temperamental at times, yes, but without doubt more alive. For you to decide ...

19 *How is a biodynamic wine analysed? Chemical analysis and sensitive crystallisation*

All modern oenology is built on the chemical understanding of grape and wine composition, as well as the transformations that take place during fermentations and *élevage* (winery operations between fermentation and bottling). This approach, which has been widespread in general agriculture for several decades, developed rather late in the wine world. Its rapid expansion began in 1956, the year the French National Diploma of Oenology (*Diplôme National d'Oenologue* or DNO) was created. The study of wine chemistry and its corresponding analytical techniques is indeed the central pillar of this university programme. The goal of these chemical analyses is to isolate some key parameters which would give enough information to explain the phenomena at work in wine, to predict them, and finally, to control them. It is what I would call 'corrective oenology', one that tries to correct the wine even before it is made, thereby moving closer to the ideal analytical parameters. As such, I find it quite interesting to note that the DNO program is traditionally taught in Pharmaceutical Departments at French universities. Is an oenologist therefore a 'wine pharmacist'?

Dear reader, I want to draw your attention here to the major risk associated with this approach. Modern oenology, with its scientific contribution, has certainly equipped the winegrower with new tools to help him make the best possible decisions during production. But very often the analysis report is given excessive importance, and by overly focusing on the necessarily simplifying numbers, one forgets to observe. The dictatorial style of the analysis report can go so far as to push some oenologists to correct everything: acidity, alcohol, nitrogen, tannins, juice to skin ratio ... I have mostly observed this trend in new world countries, but it can also happen in France. As an example, I would like to share with you an experience that really struck me at the time. I was in Argentina to carry out vinifications in one of the country's largest wineries. There were eight of us, all oenologists, who had the task of vinifying the 25 million bottles produced each year. In this very warm region, the grapes are often very ripe, but seriously lack acidity (pH=3.60 to 4.30). As the Department of Oenology teaches that 'ideal' acidity for a red wine corresponds to a pH analysis of 3.50, the oenologists tried to move closer to the maximum by adding a significant quantity of tartaric acid. In the cellar the musts were therefore systematically acidified by 1 to 2.5 g/lit. Then, one day, harvesters brought in some magnificent Merlot grapes from a cooler mountainous zone which had much higher acidity (pH=3.30). Well, the decision was made to correct them in the other direction by performing deacidification with calcium carbonate! What a shock and waste of energy for a must that could have simply been blended with one lacking acidity ... This illustrates for me the stupidity of placing such importance on numerical analyses. An 'ideal' analysis report can hide an average wine, just like a great wine can have an atypical analysis.

With respect to wine, perceptions of taste are complex,

often global. The intensity of a single parameter does not depend on its presence alone, but also on the level of other parameters. Here the principle of harmonious integration is at work. For example, in the case of the great German sweet wines, the elevated level of acidity makes their high residual sugar content more digestible and keeps them from being excessively heavy. Another more subtle example: I remember tasting a Gewürztraminer Grand Cru Goldert 2003 at Domaine Zind-Humbrecht. The wine was racy, balanced, dry and powerful and I estimated the alcohol content to be 13.5 or 14 degrees. What a surprise it was to discover that the label declared 17% volume ... which I would never have guessed as the alcohol was so harmoniously integrated thanks to the wine's high concentration. On the other hand, a wine that indicates 12 % volume but is thin, diluted and chaptalised by 1.5 degrees, will certainly produce a burning sensation on the palate.

When working with the living organisms, numbers are inevitably simplistic. Agronomists have had the same experience regarding soil fertility. For a long time they assumed that the analysis of three minerals (N: nitrogen, P: phosphorus and K: potassium) sufficed in measuring the needs of a cultivated plant. This simplistic reasoning led to the almost exclusive distribution of NPK fertilisers, which constitute the exact opposite of a healthy diet for a plant. These fertilisers, composed of soluble mineral salts, can easily be compared with fizzy soft drinks in the human diet! They sustain the plants via perfusion. Living organisms are much more complex.

Sensitive crystallisation

For this reason, I, like many biodynamic growers, distanced myself from quantitative chemical analyses a few years ago. How, then, can one evaluate the quality of a living product without using quantification? By tasting, surely, but also through the use of a technique developed by Ehrenfried Pfeiffer, and encouraged by Rudolf Steiner: sensitive crystallisation. Originally developed for the medical field (for the qualitative analysis of blood, more specifically) and agriculture, this technique aims to make the life qualities of a product visible. In practice, it involves adding a drop of wine to a salt copper chloride solution, then allowing it to evaporate in a Petri dish under constant, reproductive conditions. During the crystallisation process, the copper chloride takes on a form (like ice in a snowflake) characteristic of the vital energies of the product. Seeing the formed image, it is possible for example to observe representations as complex as they are diverse: harmony, maturity, ageing potential, oxygen sensitivity, the presence of botrytis, and many others ...

This technique is particularly powerful. In 2003, the year of the heatwave, the summer weather conditions were so extreme that harvesting began in Burgundy at the end of August. In Puligny, as elsewhere, the grapes were very ripe, but had dangerously low levels of acidity. All the oenologists continually repeated: 'We need to acidify, otherwise the wines will not age ....' At Domaine Leflaive, acidification is never practised, but 2003 was so unusual that the Technical Director, Pierre Morey, nonetheless asked himself the question. He decided to conduct an experiment: three different doses of tartaric acid were added to three samples of wine, and nothing was added to the fourth wine sample. The four unlabelled bottles were sent to Margarethe Chapelle, the oenologist who carries out sensitive crystallisations for the *domaine.* Without informing her of the adjustments that had been made, she was asked to give her opinion on the bottles. She classified the wines in descending order by energy level and ageing potential, and it was the exact order of increasing doses of added acidity! As you can imagine, this opinion reassured Pierre Morey who had decided not to acidify. A complete tasting of all 2003 Domaine Leflaive wines took place in 2011 and showed me that the acidity, of course, had not miraculously returned, but that with time, the wines had found their own balance. And, surprisingly, they had not yet reached their prime. Rather, they improved significantly after opening. They were better after three to six hours and highly oxygen-resistant. Oenology Departments teach us that the level of acidity is essential for ageing potential, and that the pH directly determines the action of SO_2 as an antioxidant, and yet...

As a conclusion to this chapter, I would like to recall what I mentioned in question 13: Does biodynamics express *terroir* better? The current division of vineyards in the Burgundian Côte and its corresponding *appellations* is essentially founded on the work done by monks during the Middle Ages. Today we have at our disposal extremely sophisticated soil analyses: the mineral composition of soil, of bedrock, and so on. We can even go so far as to measure the internal surface of clay in order to quantify its strength (a method developed by Claude and Lydia Bourguignon). Yet, every time we find a discrepancy, the only thing we can say is: 'They were not wrong!' Let us therefore put the power of chemical analyses into perspective.

Part 3

CONSUMPTION

20 *How should a biodynamic wine be tasted?*

In the first two parts of this book, I have attempted to show you to what extent a biodynamic grower seeks to produce the most living wine possible. For me, *living* expresses the opposite of *standardised*. Using this hypothesis, each tasting is a unique experience, where a special connection is created between the wine taster and the wine.

In the *standardised* view, each wine is analysed and described in a structured way. It is the realm of the left side of the brain (see question 10: Is biodynamics scientific?). It is the view of wine-tasting experts who tell you what you should smell and detect when you taste a wine. It is the style of books that categorise wines as: 'Meursault tastes like this, Puligny like that ...' It is the approach of wine critics who allocate rankings in the form of scores out of one hundred. It is a perspective on wine that has developed rapidly since the 1980s, primarily under Anglo-Saxon influence. Since then it has taken on much importance, but has now reached its limits. Indeed, this view can be applied naturally to standardised wines, corrected by modern oenology in order to obtain the best tasting scores (at the risk of mummifying them ...), but not at all to biodynamic wines. You should be aware that standardised tasting has not always been the norm. In the Middle Ages, for example, the *gourmets-experts* practised a completely different style of tasting that Jacky Rigaux explores in an excellent work: *La dégustation géo-sensorielle* (Éditions Terres en Vues, not yet available in English).

In contrast, the *living* view is the right brain's domain. It places all importance on the relationship, the unique character of the wine, the wine taster and the moment. Far from attachment to general and simplifying norms, the truth is relative, rich yet subjective, as it is based on the feelings of the taster. This point of view teaches us to taste less intellectually and to leave more room for feelings. In this regard, 'taste and feel the wine' is the title of a course developed in 2011 at the *École du Vin et des Terroirs* in Puligny-Montrachet. A few simple experiments are proposed, such as tasting blindfolded which helps to keep one's intellect quiet. The main idea is to be more conscious of the sensations felt by the entire body, which give a much more profound perception of the wine than a simple sequential description of aromas. However, this kind of less 'mental' perception is also more difficult to put into words.

Ideally, of course, everyone would fully use both approaches, the right brain and the left brain. I try to do this. After my engineering studies, solidly structured by huge doses of mathematics, physics and chemistry, I also developed my feelings through tasting, manual labour and the practice of biodynamics.

How can you in practice foster the relationship between a wine and a wine taster? It can be done quite simply by putting both of them in the best possible setting. For the wine, it is essential to create serving conditions that will allow the wine to express itself. Time plays an important role. You must take the time to taste. I am not very keen on decanting which is sometimes too aggressive. Rather, I would advise you to taste the wine after opening the bottle, to leave it for a while at cellar temperature, thirty minutes to three hours, then to come back to it. For the wine taster, it is important to be relaxed and free, without external stress. This most likely seems obvious to you because it is perhaps how you already approach tastings with friends. You should know nonetheless that these conditions are rarely in place during professional whirlwind tastings.

A particularly interesting point involves the role of the wine glass as an intermediary between the wine and the wine taster. Numerous oenologists have become interested in the question regarding glass forms and you will find an abundance of different sizes and shapes on the market, all adapted to each specific type of wine: white or red, sweet wines, Bordeaux or Burgundy, young or mature ... However, very few modern manufacturers have really studied the role of the material itself. Yet, it is the nature of the material which gives the glass its filtering properties or, in contrast, its ability to transmit information from the wine to the taster. On this subject I remember a tasting which took place in December 2009 with Bruno Quenioux at the École du Vin et des Terroirs. Among other experiments, we tasted a wine served in two different glasses. One was the Expert model from Spiegelau: tall, modern and versatile. The other one was a Baccarat model designed by Bruno Quenioux in the 1990s: small, traditional in form, made of hand blown, very clear crystal. The wine was a red Domaine de Villeneuve 2004, a Châteauneuf-du-Pape produced biodynamically. In the first glass (Spiegelau), the wine seemed powerful, full-bodied, earthy, yet excessively so. It was even a bit heavy and nearly rustic. I did not enjoy it because of its lack of elegance. In the second glass (Baccarat), the aromas were slightly less powerful, and above all, the palate still displayed some earthiness, but the rusticity gave way to a beautiful transparency, pure and crystalline, which I had not noticed at all in the first glass. I much preferred the second glass, where I was able to identify the qualities of a biodynamic wine from a great *terroir.* This anecdote is to make you aware that modern glasses, called oenological, have qualities that primarily concern olfactive analysis and often speak to the left brain. However, glasses from past centuries made of clear crystal (or perhaps even *tastevins* made of precious metals such as silver or gold), despite their rather inappropriate form hardly conducive to the aromas, certainly facilitated this subtle link between human beings and wine.

'Tasting has an escape value comparable to that of other arts. In this regard it is a source of culture as it teaches discrimination, assures judgment and reconciles us with the natural world', wrote Max Léglise. Living, biodynamic wine allows us to once again experience such tasting which might otherwise have disappeared with the proliferation of standardised wines.

I have just told you that in order to better taste a wine, it is important to be more conscious of bodily sensations felt while tasting. What does this mean? We usually learn to taste with our five senses. They provide us with clear, easy to analyse information. Sight, for example, teaches us about colour, touch, about the viscosity or texture of tannins, and taste, about acidity, bitterness, sweetness, and saltiness. The sense of smell alone is sometimes a bit more complex to analyse.

Yet you can go further and feel the effect of wine more generally in other parts of the body. You might feel, for example, the sensation of a wine 'going to the head', or, the opposite, 'going deep down into the stomach', or even, 'giving shivers down the spine'. A certain wine can make me feel lighter, more 'airy', as if I were flying. Some other wine will give me the feeling of being heavier, more anchored in the ground.

From an even more holistic perspective, some images can come to mind without going through an analysis of particular feelings. For example, 'a blazing summer afternoon under the sun on a Greek island', or even, 'a morning walk through the forest in autumn'. These images illustrate a feeling taken as a whole, which can be very complex although it does not go through the intellectualisation process. For me, it is at this level that Max Léglise's observations make complete sense. Only then does tasting match up to the other arts ...

21 *Are biodynamic wines healthier?*

Biodynamic wines are certainly healthier because biodynamics is above all a form of agriculture that focuses on the production of quality, healthy, living foods. As I have just explained above (see question 10: Is biodynamics scientific?) for Steiner, physical laws alone do not enable the understanding of the subtleties of life. There are indeed other forces which are not directly accessible by our senses, and therefore immeasurable, but which are important in agricultural work. The same understanding applies to food: a chemical composition analysis of food (carbohydrates, fats, proteins, minerals, calories, etc.) teaches us very little about their actual nutritional value, and essentially nothing about their vitality. Sensitive crystallisation, which allows one to evaluate energetic quality (see question 19: How is a biodynamic wine analysed?), shows us that biodynamic wines are on average far superior to others. The body is generally better able to digest and assimilate them. I would like to recount two experiences regarding this: one experienced personally, and the other told to me by Anne-Claude Leflaive.

My wife is a very sensitive woman. She enjoys good wine, but if she drinks even one or two glasses in the evening, she regularly wakes up in the middle of the night. Her heartbeat accelerates and it is impossible for her to go back to sleep before morning. Yet, we have noticed that with biodynamic wines, this almost never happens. It is difficult to explain scientifically, but the body tells a truth that cannot be denied.

Another example: in 2008, Anne-Claude Leflaive attended a dinner in southwest France. Among others, the famous French agronomists Claude and Lydia Bourguignon, and a former French Agricultural Minister who was initially hardly in favour of biodynamics, were in attendance. Claude Bourguignon was accompanied by a doctor friend who had been through a terrible ordeal. He was passionate about wine but, after having developed throat cancer, he could no longer drink it. With every sip of wine, the alcohol burned his throat and, to relieve it, he would swallow a glassful of antacid. Anne-Claude Leflaive asked him if he had ever tasted wines produced biodynamically. He was not familiar with them. She served him one and, for the first time, his throat did not suffer. Although incomprehensible for a Western doctor, the body had spoken.

22 *Are the qualities of biodynamic wines different from other wines?*

Today, many wine specialists claim to perceive a distinct difference between biodynamic and other wines, even organic wines. There are, of course, biodynamic growers whose opinion might well be rather subjective, but who, after all, know their wines best and have had the experience of vinifying the same parcel before and after biodynamics. There are also journalists, sommeliers, wine merchants and, of course, amateur tasters.

Tasting notes converge to say that wines produced biodynamically have more minerality and better acidity. This has also been my experience. For me, 'better acidity' refers to the quantity and also the quality of acidity, that is, riper and less green. As surprising as it may seem to you, there is acidity, and there is acidity ... And it is not chemical analysis (total acidity or pH) that allows one to differentiate between the two. For white wines as well as for red, the biggest problem a winegrower faces during harvest in warm years (or even in warm climates where the grape variety is not well adapted), is obtaining full maturity of fruit, aromas and tannins, without an overly high degree of alcohol, and especially, without losing that which provides the tension: acidity and minerality. For top white wines, there is nothing worse than having to harvest before full maturity out of fear of a substantial drop in acidity. This can happen when temperatures are very high during the days preceding harvest. This was precisely the case in Burgundy in 1997, when September was very hot. Pierre Morey and Anne-Claude Leflaive explained to me that at that moment, they realised for the first time the impact that biodynamic viticulture has on a grape's balance. The consulting oenologist who carried out analyses for Domaine Leflaive and a number of other estates in the Côte d'Or, was very surprised indeed because acidity 'had fallen flat on its face' at nearly all of his client's estates, except Domaine Leflaive! A valuable asset for producing balanced wines in a warm vintage ...

Along with the minerality drawn from the soil by the roots, tasters also describe biodynamic wines with the following expressions: *depth, verticality, anchorage.* And possibly: *length, complexity, purity*. I must admit that even if one feels the difference, the vocabulary used to describe it is not very rich, and actually quite limited. I have already explained to you the importance of feelings in tasting wines produced biodynamically (see question 20: How should a biodynamic wine be tasted?). Typically, the terms depth, anchorage, and verticality are more like attempts at expressing a feeling than precise, analytical descriptions. When there is a lack of words or descriptive language, perhaps you need to consider using other forms of expression such as poems, design or painting! Why not?

At Domaine Leflaive, biodynamics was progressively developed over an eight-year period. From 1991 to 1997, Anne-Claude Leflaive was able to carry out separate bottlings in order to compare the influence of a specific form of viticulture within the same parcel (Bâtard-Montrachet and Puligny-Montrachet 1er cru le Clavoillon). Tasting notes from this experiment were published in the *Revue du Vin de France* (in the 1990s, and then more recently in number 530 in April, 2009). Yet, the most important aspect for me is that every time I tasted these bottles blindly and without knowing which ones were biodynamic, I became aware of the importance of the global images that sprang into my mind from the first moment of contact with the wine. They allowed me to recognise the form of viticulture with near certainty. Thereafter, once I used my intellect to try to justify my choice with a rigorous analytical tasting of the different aromas and the palate, I started to get lost and was wrong approximately fifty percent of the time.

In conclusion, wines produced biodynamically generally have organoleptic qualities that distinguish them from wines conventionally produced. But even more importantly, they offer us new dimensions in wine tasting, and that excites me.

23 *How should the quality of biodynamic wines be evaluated? The role of wine critics*

In question 20: How should a biodynamic wine be tasted?, I underlined the importance of not doing a standardised tasting, but rather, fully using your feelings. I also insisted on the importance of creating conditions that allow the relationship between the wine and the taster to establish itself. Yet, too often these conditions are far from being united when it comes to professional wine critic tastings.

In the past, journalists regularly visited the growers who had produced the wines for which they wrote their tasting notes. This gave them an in-depth knowledge of the producers, their personality, their work and their history. Today, fewer and fewer journalists take the time to do this exercise, especially those who annually publish reference guides for amateurs. This is completely understandable as the number of quality producers does not cease to grow, and it would represent a considerable task. Today, they prefer to proceed with a call for samples often organised by the various wine councils. Then, these samples are tasted at lightning speed in a laboratory, a meeting room, or even a hotel room. These are sometimes real marathons where 100 to 200 bottles can be evaluated in one day.

Under these conditions, how can one not practise a standardised tasting? Even more, most journalists push the standardisation so far as to condense their impressions into one numerical score: points out of 100 for the Anglo-Saxons such as Robert Parker or the *Wine Spectator,* and points out of 20 or 10 for the French. What brutality for the wine! This does not suit biodynamic wines at all.

In addition, these sessions are generally blind tastings. It is a sales argument for guide books as they boast to their potential buyers about the critic's impeccable objectivity. Blind tasting is supposed to be more fair as all wines are assessed under identical conditions, and the taster, especially, cannot be influenced by the label. For me, this idea is essentially wrong, and in fact, I find blind tasting often very unfair. I would like to make two comments on this subject. First, wines are rarely at the same stage of development at the same time. An example: some wines are made to be drunk early, others to be stored. How can a journalist take this into consideration if he does not have access to such basic information as the type and length of ageing a wine underwent before bottling, or even the date of bottling if very recent? It is unfair. All this information would be available if a journalist went and tasted the wine in the producer's cellar. Second, I believe that any critic deserving of the name must have the necessary experience to properly use the information expressed by knowing which wine is being tasted so that their judgment is not biased. No serious professional would invent imaginary qualities of a wine based on its reputation alone. As far as I am concerned, I would tend to be even more demanding with a 'great label' as consumer expectations, and most likely its selling price, justify it.

Blind tasting is only of interest for those looking to explore their own taste. How fascinating to try to identify an appellation, a vintage, or a producer, just like the tasting competitions among the best sommeliers. If you are passionate about this, I have only one piece of advice for you: try tasting completely blind, either in the dark or with your eyes blindfolded. You will realise what tremendous influence the sense of sight has on our mental processes. I wonder when we will see the first guide to wines tasted blindfold, or even a guide that rejects the standardising dictatorship of numerical scores?

24 *Are there 'bad' organic wines?*

I have often heard it said, 'For a long time organic wines were generally worse than other wines, some frankly undrinkable. This is the reason why organic wines, in the past, did not have an image of quality with tasters.' To be truthful, given my age, I was not there to taste them. It is therefore difficult for me to say if that was indeed the case, or rather, if it goes back to a time when organic wines were simply too revolutionary and so their taste, not being the norm, was more shocking than it is today.

There are certainly extreme organic wines that the current norm qualifies as 'faulty': oxidation, cloudiness, microbiological contamination, even refermentation in bottle, high volatile acidity (when the wine comes dangerously close to turning into vinegar). Before making any comment, let us note that there are probably just as many faulty wines produced by conventional viticulture.

Yet what really annoyed a number of professionals, journalists in particular, was to imagine how people could rave about a wine obviously displaying such flaws, and forgive it simply because it was organic. As an American friend of mine would say: 'There is an overwhelming opinion today that too many organic producers purposely make bad wine because they say that is what organic is. Too many winemakers believe they should do nothing in the cellar and the wines suffer.' At least one French journalist became so angry as to talk about 'organic-idiots', condemning fanatics who had lost all sense of judgment and blindly followed the organic bandwagon.

Such language of condemnation was strong and in his rage, the writer presumably did not see the real change at work: a reorientation of wine lovers' own personal tastes, so redefining the norms that separate quality from fault. In any case, there was a great wave of enthusiasm on the part of wine lovers, professional wine merchants and sommeliers, supporting these organic producers in their pursuit of authenticity. Thus, consumer wine appreciation is becoming more personal and it is up to each individual to determine what they like and dislike in a wine. Then, in the end, every individual has to decide for themselves if they find satisfaction or a sense of well-being when consuming the wine.

I admit that there was also a time when, with great pleasure and tremendous curiosity, I would taste wines that were sometimes truly extreme. Today, I wholeheartedly accept certain faults, aromatic in particular, provided that I feel an authentic soul (personality, vitality, energy) that transmits emotions to me.

Now is it my turn to ask you a question: in your pleasure-seeking search through the world of wine, what really gives you enjoyment? Does your enjoyment honestly lie in the idealistic pursuit of near-perfection, a wine without fault but with the risk of being lifeless? Isn't such enjoyment a bit too cerebral, encouraged by modern society with its magazines endlessly displaying pictures of models perfected by plastic surgery, and retouched by Adobe Photoshop? Or, on the other hand, does your enjoyment stem from a real encounter with a product free of any covering up? This is what fairy tales (such as *'The Ass's Skin'* by Charles Perrault) tell us. By looking beyond surface imperfections, those who make the effort can arrive at a wonderful discovery. Is it possible to discover a prince or princess behind the appearance of an 'organic idiot'?

25 *Are biodynamic wines unpredictable?*

As I already had the opportunity to explain in question 18 (Do biodynamic wines age better?) I am convinced that wines produced biodynamically are generally more alive, and therefore more sensitive than wines produced conventionally. For me, more sensitive means more receptive to external influences such as the seasons, the weather, the moon, the place, the people, and so on.

Is this a quality or a fault? Does 'more sensitive' also mean more fragile? It is up to you to form your own opinion. Even if the social norm in France has, for a long time, confirmed the contrary, as far as I am concerned, sensitivity is a great quality. This could be said for human beings as well as for wines!

At this point I would like to go into detail on one aspect of sensitivity. In general, when you hear about a moody person, especially when it has to do with a woman, reference to the moon or to the lunar cycle is never far away. Don't we talk about someone being a 'lunatic'? So how does wine react to the lunar cycle? Intrigued by this question, over the course of one year I paid particularly close attention when tasting the 2007 vintage during its ageing at Domaine Leflaive. Nearly every day, sometimes twice a day, with various clients I tasted all the wines from the vat. Then, I routinely consulted the sowing calendar (biodynamic lunar calendar) to compare it with my tasting assessments. It was particularly interesting to conduct this experiment with wines in vat for two reasons: Firstly, the wines are much more free and alive in vat than in bottle where they are confined. Secondly, in a vat of over 2,000 litres I could taste the same wine from the same container every day, whereas wine in a bottle that has been open for several days evolves quickly upon contact with air. And, if I open another bottle, there is already a difference as it is not exactly the same container.

Undoubtedly you are already familiar with one of the biodynamic methods for describing the influence of the moon. There are four types of day: *Fruit, Root, Flower, Leaf* (see question 9: What is the planting calendar and what is its purpose?). These correspond to each element (Fire, Earth, Wind, Water) of the constellation in front of which is the moon, which allows a winegrower to focus their agricultural work on the cultivated plant itself. The calendar is based on work by the German biodynamic researcher Maria Thun (who died in 2012) and her son, Matthias Thun (*The Biodynamic Sowing and Planting Calendar,* Floris Books). A small booklet also exists in English which uses these results for wine tasting *(When wine tastes best: a biodynamic calendar for wine lovers,* Floris Books) and even, more recently, an iPhone application (*Wine tonight?*). The booklet explains that *Fruit* and *Flower* days are best for wine and that *Root* and *Leaf* days should be avoided. It is a bit too simplistic and the following is what I have been able to personally observe. The calendar days do not always have a noticeable influence on tasting, but every time I noticed a significant influence, it was always in line with the lunar calendar.

Fruit day: the vine is grown for its fruit and a *Fruit* day is naturally the most favourable day for work in the vineyard and the winery, as well as for tasting. The wine is generally very expressive and open aromatically, with a harmonious balance on the palate.

Root day: this type of day was found to be rather unfavourable as the wine can show itself to be a bit muted, as it is rather closed aromatically, particularly on the nose. In the palate the wine can also be quite closed and reveal the structure. Thus it may seem austere and hard. But for wines from great *terroir* which must naturally have good structure, it is a type of day that can be interesting as the minerality is very discernible.

Flower day: this type of day is generally favourable to tasting as the wine is very expressive, sometimes even more open on the nose than on a *Fruit* day. You should nonetheless be cautious as I have noticed that *Flower* days tend to exacerbate whatever is delicate and volatile: certainly the aromas, but also alcohol and volatile acidity (a sharp odour similar to that of vinegar). This seems especially to affect very rich wines, sweet wines, or those having undergone very long barrel maturation.

Leaf day: this type of day is usually unfavourable to tasting, and should even be avoided altogether as it only displays unpleasant elements in the aromas and the structure of the wine. Such elements evoke vegetative matter: vegetal on the nose and a lack of elegance, hardened acidity and the impression of underripeness. For rich wines with low acidity, there can be a bitter sensation, as well as harsh tannins for reds.

Each lunar influence generally lasts two or three consecutive days, and there appears to be a cumulative effect. A *Flower* influence, for example, is much more evident after two days than at the beginning of the period (the initial hours) when the influence from the preceding period can still be felt.

Observing lunar influence is fascinating, but we should not become overly dependent on it. In organising a successful tasting this is only one aspect of the environment among many to be taken into account: serving temperature, the general atmosphere, the type of food served, guests' tastes, the personality of the *terroir,* vintage characteristics ... Rather than classifying the days as good or bad, you should grasp the subtle interactions among all the elements. This is what constitutes the know-how of a great sommelier. Regarding the wines of Puligny-Montrachet, for example, a *Root* day suits a vintage such as 2006 very well: a rich, opulent, expressive vintage, with low acidity. On the other hand, for a *Flower* day, I would choose a wine from the 2007 vintage: cool vintage, very tight, good minerality, and somewhat closed in its youth.

And so one can say: this wonderful sensitivity is lost today in modern, overly rigid wines produced by chemical agriculture, but reclaimed in biodynamically produced wines. This sensitivity has always existed in wine. As proof, let me offer you the following quote from Jean Carmet referring to Charles Joguet, the great winegrower from the Loire. It was told to me by Bruno Quenioux, a wine merchant and taster, and goes: 'I have sometimes seen wines disappear in the presence of an intruder in the cellar and reappear once he left. Let us know therefore what happens in the mind of a wine that resembles the man who made it!'

26 *Are biodynamic wines more expensive?*

The notion that wines produced biodynamically are necessarily more expensive than other wines because of the additional constraints that biodynamics imposes, is relatively common. Well, in my opinion, the reality is a bit more complex. In this section I hope to shed light on the subject, both for you as a wine lover, and for producers who hesitate to change their methods because of the additional costs involved.

First of all, let us look into cultivation costs. It is clear that biodynamics generally entails an increase in working hours and therefore, increased labour costs. This refers most specifically to the time spent meticulously carrying out the various grape-growing tasks: ploughing (possibly even with horses), tasks done during vine growth (disbudding, and so on), manual harvesting; and, above all, work specific to biodynamics: the preparation of dynamisations and the spraying of the preparations and infusions, the spreading of prepared compost, observance of lunar rythms, and 'sensitive' observance of one's vines. The former tasks can be very time consuming and generally depend on how demanding the winegrower is. Yet, they are not at all exclusive to biodynamics. With respect to the latter tasks, which are the only additional biodynamic work, the extra time spent is not really significant, and I would estimate it to be between ten and twenty hours per hectare, annually. It essentially involves organisational issues such as working early in the morning or late in the evening to spread the preparations. In general, most of this slight cost increase is offset by a significant reduction in the purchasing costs of agrochemical products. Insecticides, herbicides, systemic fungicides and other anti-rot products are actually very expensive compared to sulphur, copper and natural plants. Collecting a bag of fresh nettle in the spring is not very costly! For me, I am very happy if, for the same price, I can spend less on chemical products and machines, and more on human labour.

Before leaving the subject of production costs, I would like to address the issue of yields. It is clear that, in the short-term, yield directly determines the profitability of an agricultural enterprise. The cost of producing 30 hectolitres (4,000 bottles), 60 hectolitres (8,000 bottles) or 90 hectolitres of wine (12,000 bottles) from one parcel of vines is essentially the same. The quality will, of course, be very different in each of the three cases. This depends on the choices made by the grower, and on the rigorously regulated statutes of the French *appellation contrôlée* system. Yet, I must say that biodynamics can certainly be applied whatever the chosen yield level. It is absolutely wrong to think that it would automatically lead to a reduction in yields. The biodynamic grower has at his disposal preparations which aim to increase vigour and therefore grape production, and other preparations that decrease it.

The following quote from the director of a large estate in Burgundy, based in Beaune, reflects my point of view: 'Contrary to preconceived ideas, applying biodynamic methods leads neither to a significant reduction in yields, nor to any noticeable increase in production costs, compared to conventional agriculture'.

I would like to make here a general comment about the problems associated with reduction of costs through the implementation in agriculture of the industrial model. During the first years of my professional career, I devoted myself to optimising the production costs of supply and distribution, primarily in large factories. I worked as an organisational consultant in various industrial sectors: automobile, petrol, pharmaceutical, and animal feed. With the industrial model, the pursuit of cost optimisation (based on the theory of monetary scarcity) initially involves the standardisation of processes and products ('fordism' in the automotive industry at the beginning of the twentieth century). The standardisation of processes also imposes the standardisation of raw materials. Otherwise, how could a unique and consistent formula be maintained if different raw materials were used? Thus, dear reader, it is particularly important to be aware of this key point: agricultural commodities are naturally variable because they come from living organisms. 'Living' means 'diverse', but certainly not 'standardised' to which it is opposed. For me, this is the fundamental reason for which the industrial model's way of reasoning (cost optimisation through process standardisation) is a denial of life and cannot apply itself sustainably to agriculture. Denial of this evidence pulls agriculture into a vicious circle that usually leads rather quickly to a significant loss of quality, and sometimes later, the opposite of the desired goals: a decrease in production linked to a decrease in soil fertility, and an increase in costs from the purchase of agrochemical products.

First example: the decision to consolidate land in favour of mechanisation and productivity. In the end, it was realised (in Europe) a bit too late that the destruction of hedges leads, on one hand, to a visible decrease in the biodiversity of fauna and flora (birds and insects in particular), and rapid soil erosion through the loss of clay on the other. This second point is critical because it leads to a loss of soil fertility within a few decades. As a result, many specialists foresee the risk of a significant reduction in the land's productivity within the next ten to twenty years.

Second example: the standardisation of crops which is a denial of the notion of *terroir.* One talks about vine *terroir,* but there is also wheat *terroir,* asparagus *terroir,* carrot *terroir,* and so on. Because of pressure from the animal feed industry, we can witness aberrations such as generalised corn production in Alsace, a region in northeast France. Corn, however, requires a great deal of water and Alsace is a region that receives very little rain because of the Vosges mountains which act as a barrier (it rains as little in Colmar, in Alsace, as in Montpellier, in southern France). The result: a visible aggravation of drought conditions.

Fortunately, viticulture is *the* form of agricultural production which has been able to resist this general trend and defend the diversity of its products through the diversity of its *terroirs.* End of my general comment.

Let us go back to the question of wine prices. Price is one thing, but for many wines, the selling price depends mostly on quality (real or imaginary quality, some might say). Let us talk more generally about the wine's image with the consumer. This image encompasses quality, history, scarcity, fashion, and certainly many other factors. In viticulture, an estate's or an appellation's profitability builds itself less on the reduction of its production costs at any rate, and more on quality recognition and image of its wines. On this matter, I much prefer to buy an excellent Clos Puy Arnaud from Thierry Valette who makes an exciting, elegant wine full of feeling from an excellent limestone *terroir* composed of starfish fossils, a wine that is produced biodynamically (but classified *Côtes-de-Castillon*), than a wine from some of his *Saint-Émilion Grand Cru* neighbours who have nothing grand but their name and are yet sold at twice the price. But this stays between you and me. Regarding a wine's selling price, I find myself in an awkward situation. On one hand, I do not want biodynamics to be reserved for the élite, but rather to be available for all price levels. It is part of the product's nutritional quality before its organoleptic quality, and should be accessible to all. On the other hand, I also hope that the consumer will discern the value of biodynamic wines. And, the fact that they are more sought after than other wines will certainly be reflected in the price.

In conclusion, no, biodynamics is not a form of agriculture for the rich. There are a number of 'simple wines' at low prices. In the village next to mine, Didier Montchovet produces an excellent *Bourgogne Grand Ordinaire* red for 5 euros per bottle. In another region, Guy Bossard's *Muscadet* (produced biodynamically, with draught horses) is an excellent reserve wine at a price of 6.80 euros. Yet, it is clear that for growers cultivating the best *terroirs* and pursuing excellence, biodynamics is a means of going even further. Quality improvement justifies the fact that their wines are more sought after, and their value can indeed be a reflection of this.

Part 4

THE HISTORICAL AND PHILOSOPHICAL CONTEXT

It is common practice to introduce a subject by situating it in its proper context. However, in the case of this introduction to biodynamics, I have deliberately placed it at the end in order to reserve the first part for the most practical questions, those which concern you directly as a wine lover. So this fourth part is addressed to those who would like to know more about the historical and cultural background. Here I will talk less about the vine and wine, and more about an understanding of the living world, a vision of the universe, and even philosophy.

Some of these ideas or advanced theories may surprise you or even shock you. Yet, mentioning them here allows you to situate biodynamic practices in a larger scheme, and to discern their reasoning more clearly.

27 *Who was Rudolf Steiner?*

Rudolf Steiner was a German-speaking spiritual philosopher. Born on February 25, 1861, in Kraljevec, Croatia, which was part of the Austro-Hungarian Empire at the time, he was Austrian and his father was employed by the railway. During his four years as a science student (mathematics, chemistry, natural history ...) at the Technical University of Vienna, he became interested in philosophy, first with Kant, then Goethe, at a very early stage. He then interrupted his science studies to devote himself fully to philosophy. His first job was to carry out the publication and commentary of scientific works by Goethe. Rudolf Steiner later moved to Germany, first to Weimar, then to Berlin, and eventually obtained his PhD in Philosophy at the University of Rostock. After 1900, he joined the Theosophical Society where he met the Society's President, Annie Besant, as well as Marie von Sievers whom he eventually married. He served as general secretary of the German Theosophical Society from 1902 until 1913. At that time he founded his own movement, the Anthroposophic Society, in Dornach, Switzerland, near Basel, to develop and spread his spiritual philosophy: anthroposophy.

He devoted the rest of his life to applying this philosophy to various practical areas of human life in order to inspire them with new ideas: art, architecture, education, medicine, religion, politics, economics, and ... agriculture. He gave courses and over six thousand lectures. This spirit of reform is particularly recognised in the education sector. Heiner Ullrich of UNESCO writes:

> The revolutionary mood in a defeated Germany in 1918 and 1919 brought Steiner the opportunity to try out his ideas on education in a new school. On September 7, 1919, he ceremonially opened the first Independent Waldorf School as a combined co-educational primary and secondary school for 256 children drawn mainly from families of workers at the Waldorf-Astoria cigarette factory in Stuttgart, Germany. (UNESCO International Bureau of Education, 2000)

In June 1924, Steiner gave a series of eight lectures in Koberwitz known as the Agriculture Course. It is precisely this Agriculture Course which is the foundation of biodynamic agriculture. He died the following year on March 30, 1925.

28 *What is anthroposophy?*

Although the purpose of this work is not to be a book on anthroposophy, I believe it is important to provide you with some insight into the subject. This will undoubtedly allow you to more easily grasp the rationale in which biodynamic practices were developed.

First, let us look at the etymology of the term chosen by Rudolf Steiner: from Greek *anthrôpos,* man, and *sophia,* wisdom. It is indeed a philosophy centred on mankind, or stemming from mankind, to explain the world. Second, I offer you, though brief, I admit, the definition from the French dictionary, *Le Robert:*

> Anthroposophy: doctrine established by Rudolf Steiner. Similar to the thinking of Goethe who strongly influenced him, Steiner's doctrine seeks 'a path of knowledge aiming to guide the spiritual element in the human being to the spiritual in the universe.' Thus, going beyond the exclusively technical, materialist and destructive character of modern science which, since Kant, refuses to recognise the true place of human beings, anthroposophy offers an understanding of human nature capable of giving the individual his true place at the heart of the universe, 'to expand and deepen our sense of social, pedagogical and medical activity.'

What does this mean in practice?

The following is the first important idea of Steiner: living beings are not composed of matter alone. A living organism is matter plus a certain form of energy. It is the latter which distinguishes, for example, a stone from a plant or an animal. A living organism is not governed by material laws alone (physics, chemistry), and is able to regenerate, or reproduce. Yet, this 'energy' that characterises life is not visible to the eye.

This leads us to Steiner's second major idea regarding the *sensible* world (or material world) as our five senses perceive it and materialist science describes it. However, this is not the only world. There is also a world that Steiner calls *supersensible* (or spiritual), not directly perceptible by our senses, and which he said could be described by a science yet to be developed: *spiritual science*. This is the subject of his book, *Esoteric Science,* published in 1910. For Rudolf Steiner, this supersensible world was, for a long time, the realm of beliefs, often governed by religion. Yet, he considered that after the Age of Enlightenment and the tremendous development of rationalism, it was time for mankind to move on to a new phase. Every individual must find their own path to accessing the supersensible world, relying no longer on priests, magicians, or dogmas.

In *Esoteric Science,* Rudolf Steiner formalises a rational approach of the understanding of the supersensible world. He wrote:

> The important point is that [this book] attempts to see into spiritual worlds by using means that are both possible and suitable for souls at this present stage of evolution, and that it considers the riddles of human destiny and of human existence beyond the limits of

> birth and death by these means. The point is not that this attempt bears some ancient name or other, but that it is aiming at the truth. (p. 430)

This spiritual science, as he calls it, bears the name 'science' as it involves an approach based on precise and reasoned knowledge of the subject one studies. However, contrary to material science which utilises conceptual thinking exclusively (the left brain), Steiner places importance on 'imaginations' and 'inspirations' (the right brain). More than proof obtained from causal links, he prefers the truth of individual, personal experience. His tools of investigation are human beings themselves, therefore, *anthroposophy.*

29 *What was the state of European agriculture when biodynamics was established?*

I have often heard people talk about the 'ravages of agriculture over the last fifty years'. However, you should know that the problem is much older, even if it has certainly accelerated and become more widespread over this time period. Some individuals even became aware of it as early as the 1920s. According to Ehrenfried Pfeiffer, farmers who witnessed the damaging effects of modern agricultural techniques, sought advice from Rudolf Steiner in such terms: How can one stop the degeneration of seeds and nutritional value? (Reported by E. Pfeiffer in the appendix of the Agriculture Course.) For Steiner the problem was very clear. In 1924 he wrote, '...over the course of the last decades one has seen a degeneration of all agricultural products with which man nourishes himself, a degeneration at an extremely rapid pace.'

The end of the First World War in 1918 marked the introduction of two significant types of product in agriculture. The following emerged from the reconversion of arms factories: ammonium nitrate (nitrogen fertilisers) and organophosphates (derived from chemical weapons and the basis of powerful insecticides). From a symbolic point of view, it is staggering to think that a large part of agrochemistry is built on knowledge generated for military purposes with the aim of inflicting death!

The origins of this trend are undoubtedly to be found in the previous century, during the era known as the *agricultural revolution,* then, during the *industrial revolution* which followed. In this regard, viticultural history is emblematic as the primary viticultural pathogens first appeared in the middle of the nineteenth century. Before that time it was possible to grow vines without purchasing treatment products from a retailer. A few dates:

1847: the arrival of powdery mildew in France ('came from North America' it seems ...). 1863: the arrival of phylloxera (again, 'from North America'). Thousand year-old French vineyards were essentially decimated in less than twenty years!
1878: the arrival of mildew (from ... North America).

Still today, mildew and powdery mildew are the two vine diseases that growers fear most. One and a half centuries of chemical struggle have not been very conclusive. This agricultural revolution has witnessed the arrival of new techniques and innovations only to be followed a few decades later by new diseases! Could there be a connection?

I would like to draw your attention to a more profound question. Apart from new methods, the agricultural and industrial revolutions have led to an upheaval in attitudes, and the status of the 'peasant' has changed completely. This period has witnessed the disappearance of *commonsense farming,* in favour of *progress* brought about by materialist science. The following quote which really touches me comes from an interview on this topic by François Bouchet who was the first biodynamic consultant for viticulture. He talks about his early days as a winegrower in the 1950s. Concerning traditional peasantry, he writes:

> People would tell me: 'Me, I do it like this, but I cannot explain why. You, you went to school, therefore you know better than me.' This farmer's modesty touched me profoundly and I did everything possible to save this lost knowledge.

And yet, he adds:

> Contrary to what one thinks, they did not have the slightest notion of ecological awareness. They did not know they were poisoning nature. They did not know that they were poisoning themselves ... And that was extremely distressing given that no one had informed them about these types of things. They spread fertilisers because 'those who knew' told them to do so. They were reticent about all of it, but could not formulate it. (Bouchet 2005)

And I will add to these sentiments, Steiner's own observation which can be found on the last page of the Agriculture Course:

> I have always found science to be extremely stupid. Thus, in order to make this science more intelligent, we ... are trying to bring some 'peasant stupidity' into it. Then this stupidity will become wisdom in the eyes of God.

30 *What does the word 'biodynamics' mean?*

When Rudolf Steiner gave his Agriculture Course in 1924, he did not use the term 'biodynamics', and probably did not between this date and 1925, the year he died. The term was employed for the first time after his death. In the years following his lectures in Koberwitz, the Agriculture Course spread very little. Steiner himself had hoped that his course would remain confidential for a few years, four years at least. This corresponds to the minimum time frame necessary to put the new principles into practice, and to obtain the first credible results. During this time, only a small circle of anthroposophic farmers in Koberwitz were able to dedicate themselves to the new task: they also worked in collaboration with the scientific section of the Goetheanum in Dornach. Only farmers were involved, because for Steiner, everything concerning the application of anthroposophy in agriculture had to be evaluated in its most practical aspects, and especially not at the level of ideas and philosophy alone. He wrote:

> One condition for success, however, was strongly and repeatedly emphasised: for the time being, the content of the course must remain the spiritual property of the Circle of practicing farmers. Although some people only casually interested in agriculture were also present

> at the course, they were not permitted to join the Circle ... These things will only be able to live up to their true potential if the content of the course ... is tested by the farmers. Some things will require four years to try out. In the meantime the practical pointers that were given are not supposed to stray outside the agricultural community. These things are meant to enter right into practical life, so it does not good just to talk about them.

The collection of lecture transcripts was initially entitled *Biological Fertilisation,* which is to say, how to nourish the soil while respecting the principles of life. Then, around 1930, it became *Biological and dynamic agriculture,* which emphasised the significant contribution of this method: a way of farming that understands and works with the forces whose balanced expression allows for healthy plant and animal growth (dynamic, as we saw earlier, from the Greek *dynamis,* force). Soon afterwards the expression was contracted to *biodynamic agriculture.*

Although biodynamic agriculture was established by the initial prompting of Rudolf Steiner, it was largely completed after his death through research and experiments carried out by his collaborators and their successors. As we will see later on (see question 31: What is the content of the *Agriculture Course?),* Steiner's practical contribution lies essentially in the development of preparations to be used as a complement to organic fertilisation. The function of these preparations is to reactivate the cosmic forces from which the soil and plants have been cut off by modern agricultural practices. He also proposed techniques for the regulation of weeds and harmful insects or animals through incineration. But their use is rather limited in practice. Among the researchers who succeeded him, I would cite Ehrenfried Pfeiffer who developed the sensitive crystallisation technique which enables progress to be observed. Maria Thun made a tremendous contribution to the understanding of cosmic influences, and as we saw earlier, her work led to the publication of the famous *Sowing and Planting Calendar*. She also created a new preparation known as CCP, or *Cow Pat Pit Compost*. Alex Podolinsky as well, created another preparation: 500 P.

31 *What is the content of the Agriculture Course?*

The *Agriculture Course* is the written transcript from notes taken in shorthand of the content of eight lectures given by Rudolf Steiner, June 7–16, 1924, in Koberwitz. It also includes the question and answer sessions which followed each lecture as well as the report drafted by Steiner on June 20 upon his return to Dornach. It is not therefore a farming manual as such, drafted and finalised. It is important to keep this point in mind in order to maintain a critical perspective if you decide to read the *Agriculture Course.* Steiner himself took the time to write a note to the reader of his published lectures insisting on this fact:

> The content of these publications was intended for oral communication, not for print ... One needs to simply accept the fact that in these shorthand reports which I have not looked through, mistakes can be found.

At this point I will detail the major ideas established by Rudolf Steiner over the course of each of the eight lectures. A thorough presentation and explanation of their content would require a complete book. Thus, in the following, I will simply bring to light some key ideas.

First Lecture

For Steiner, agriculture finds itself in a dead-end street with the ubiquitous presence of materialist scientific reasoning, the loss of farmer instinct and, the widespread use of chemical fertilisers and industrial methods. The anthroposophic approach, or spiritual science, can engender the necessary broadening of thinking in the way one regards plants, animals and soil. First of all, when growing plants, one must take into consideration the influence of the cosmos as a whole, for the cosmos (sun, moon, planets and distant stars) influences life. Modern humans have partially distanced themselves from this awareness, but this connection remains particularly important for plant life. Thus, a plant's growth is dependent on a field of cosmic forces which, although undetectable by the five senses, is nonetheless real. Steiner compares it to a magnetic field that positions the needle of a compass: 'It would be ridiculous to try to explain the behaviour of the compass needle by looking for the cause in the needle itself' (p. 16). On the contrary, we must broaden our perspective and become mindful of all the terrestrial influence acting through the medium of its magnetic field.

It is also necessary to change one's view on geology and the manner in which it influences life. For Steiner, cosmic and earthly forces are propagated and transmitted to the plant via two minerals of opposite polarity: silica and limestone. Limestone is related to the forces of growth and reproduction. Silica, on the other hand, which constitutes more than one quarter of the earth's crust, is the carrier of distant cosmic forces, the forces of structure and limitation. We can train ourselves to observe how the relative proportion of these two poles manifests in the forms and colours of various plants. Clay facilitates and guarantees exchanges between these two opposing forces. This is the principle of *threefolding* which was dear to Steiner: two polarities and one mediator, which can be found in all of his works (health, economics, education), but which I will not develop here.

Second lecture

Steiner again examines the role of silica and calcium (limestone is mostly calcium carbonate) and develops it in connection with planets and cosmic forces. He presents two fundamental ideas. The first is the notion of *agricultural individuality* or *agricultural organism*. He believes that soil, of a farm in particular, can be compared to a self-contained individual entity and should *aim to become a self-sufficient individuality*. Stated more clearly, a healthy farm should be able to produce everything it needs on its own. It is in this light that animals play a crucial role, notably via the manure they provide which, in turn, fertilises the soil producing the plants that nourished the animal.

Secondly, a plant is most receptive to cosmic forces at the time of germination. Thus, within the seed a 'mini-chaos' is produced during which the current cosmic forces leave their imprint on the future plant that will develop. This is also the role of biodynamic dynamisation which creates rhythmic chaos, especially useful for plants that are reproduced by vegetative growth (the vine, for example), and thus do not regularly go through the seeding stage. In this case, dynamisation facilitates a special reconnection to the cosmos (see question 12: What is dynamisation?).

Third lecture

In this lecture Rudolf Steiner gives a real course in biochemistry from a spiritual science point of view. He takes the principal elements: carbon, nitrogen, oxygen, hydrogen, and sulphur, and examines each element's role in life. Carbon, the building block, the structure of the organic material world, sustains the process of incarnating what is spiritual (the 'ideal' world) in matter. Sulphur is the sculptor, carbon's architect. Oxygen enables the vitalisation of carbon and instils in it ethereal forces (those responsible for vegetative growth). Nitrogen is the carrier of astral forces (those which distinguish animals from plants). Hydrogen, the lightest element, is the mediator. I will not explore these fascinating questions in detail here, but the new perspective they offer on agronomy allows one to view current growing practices from a completely different point of view.

Fourth lecture

This lecture deals with the management of manure and compost. Steiner explains that the conventional approach, which thinks in terms of quantitative analysis, is not adequate. The essential point of fertilisation is to compensate for the *living forces* drawn from the soil by the plant, and exported during harvest. Yet, the commonly used mineral fertilisers are devoid of life forces. By definition, mineral is not alive and will never increase the life forces of a soil. Mineral fertilisers do the opposite. Only organic fertilisers can provide plant or animal life forces. It is advisable therefore to pay close attention to how compost is prepared in order to preserve the most forces possible.

Steiner then gives the formula for two useful preparations

to optimise the effect of manure. The two act on the two complementary forces described in the first lecture. One activates the effects of growth forces, and the other, the effects of structuring forces. The first, known as cowhorn manure, or later, preparation 500, is a mini-compost obtained from manure in a cow horn, matured in the soil, then dynamised. The second, known as cowhorn silica or preparation 501, is a cowhorn filled with very finely ground silica which undergoes a similar process.

Fifth lecture

Steiner concludes the topic of manure by describing in practical terms how to prepare the compost. He gives a list of six other preparations to be incorporated into it in order to activate all the forces linked to the principal minerals. 'It is not a question of merely augmenting the manure with substances that we believe will be of benefit to the plants. It is a question of infusing the manure with living forces' (pp. 92f). These additions are: yarrow for the potassium process (preparation 502), chamomile for the calcium process (503), stinging nettle for the iron process (504), oak bark also for the calcium process (505), dandelion for the silica process (506), and valerian flowers for the phosphorus process (507).

Sixth lecture

In this lecture, Steiner examines in more detail the theme of plant life and diseases, from which agriculture suffers. The first idea is that, contrary to animals and humans, a plant cannot fall ill by itself. Its disease can only be the reflection of an imbalance in its environment.

Plant life depends on external influences and, in particular, on the two significant forces, complementary yet opposing in a sense, and described in the first lecture. On one hand, growth and reproductive forces are Earth forces, strongly stimulated by the moon, especially in the presence of water. On the other hand, structuring forces and fructification forces (fructification is synonymous with the end of growth for most plants, and it is for this reason that fruit is so rich in life forces) are under distant cosmic influence (Mars, Jupiter, Saturn, and so on). Where cultivated plants are concerned, health is entirely a question of how the farmer succeeds in ensuring the proper balance between these two types of force. If, for example, terrestrial forces are dominant due to an intense moon (full moon) and excessive water in the soil (heavy precipitation), 'As a result, the seed, or the upper part of the plant, becomes a kind of soil for other organisms. Parasites and all kinds of fungi appear' (p. 128). Here, Steiner offers one of the fundamental aspects of biodynamic reasoning: in order to prevent a number of plant diseases, it is often necessary to restore balance where an excess of growth forces exists. This can be done either by stimulating limiting cosmic forces, or by consuming excessive growth forces. He cites the use of horsetail infusion as an example. Given this understanding of disease, the modern fungicide battle is useless and doomed to failure in the long-term.

Seventh lecture

The seventh lecture is truly a course in environmental ecology, viewed, however, from an energetic perspective, that is, from a point of view of the balance of forces described above. Steiner begins by reminding his audience that in nature, all things are in mutual interaction. Consequently, one cannot think rationally about a plant or a parcel without taking into account its immediate and more distant environments. He explains the subtle relationships that exist between fields, orchards, pastures and forests, as well as the role of insects, earthworms, and birds, among others. One example is that of birds that aid in the distribution of astral forces, from forest to field in particular. He writes, 'That is how a real division of labor between the birds and the butterflies came about in nature. These winged creatures work together in quite a wonderful way, sharing the work of distributing the astrality wherever it is needed in the air above the Earth's surface.' What is true above the ground is true below the ground, as with earthworms and larvae, for example.

Eighth lecture

Rudolf Steiner ends his lecture series by focusing on the function of food: livestock feed, and, above all, food's action and effects on the human diet. Here again, he holds a different view on these subjects which is unusual for a modern Westerner. I will let you decide.

First, Steiner reminds his audience that digestion is a process of breaking down food. For him this means separating substance from energy. Each has its function inside the body, as we will see later. It is important, therefore, for food to be *alive,* and to not only contain substance, but energy as well (which is to say that it contains earthly forces and/or cosmic forces).

Second, two complementary processes exist. The first is that of *earthly food*: that which we ingest into the digestive tract through the mouth. This digestion provides on one hand energy for our metabolic system (muscle and organ activity), and, on the other hand, the constructive matter of our neurosensory system (brain, nerves, and so on). The second process is that of *cosmic food*: that which we absorb from the air through our skin and senses. It is complementary to earthly food and provides us with the energy needed by our neurosensory system and, through a process of densification, the matter which constitutes our metabolic system. For instance, have you heard about people who claim they 'feed off light'?

Finally, according to Steiner, agriculture is fundamental as it is the basis of all human life. It determines not only the health and physical activities of humans, but also their psychological and social life. He gives an example of the widespread consumption of tomatoes and potatoes:

> [Potatoes] too act extremely independently, ... [they] enter the brain and make it independent ... From the time potatoes were first grown in Europe, excessive potato consumption has contributed towards making human beings and animals materialistic ... And that is why it is so important, that agriculture be related to the whole of social life.

32 *Are biodynamists a sect?*

Let us look at the word in its original meaning. *Sect* comes from Latin *secta*: the path followed (from the verb *sequor*: to follow). This makes me think of the concept of *'way'* in the Far East, whether in martial arts (Kendo, Aikido, Judo), art (calligraphy), or even the tea ceremony. For Orientals, it relates to a path of development of the person as a whole (physical, moral, and spiritual) from learning and mastering a technique. In this sense, biodynamics could undoubtedly be considered a 'way of agriculture'. Why not?

Yet, in France today, the word 'sect' has become pejorative. It designates a group of dogmatic individuals with obscure, esoteric beliefs which set them apart from others. So no, biodynamists are not a sect. Yet I recognise the fact that some people might misunderstand or even fear the renewed influence of such things when hearing talk of astral forces, cosmic calendars, and so on. I hope that by now, after having read this book, your understanding allows you to hear these terms without undue reactions!

In fact, in viticulture today, very few biodynamists are anthroposophists. They are, above all, country people, pragmatic farmers searching for a good way to resolve the dilemmas of modern agriculture. Rather than a sect, it is a body of free individuals, often with strong personalities. A typically Cartesian wine lover recently told me,

'Biodynamics as such does not interest me. However, I have noticed that the growers who are passionate about biodynamics have unique and fascinating personalities, and that does interest me!'

33 *Did Rudolf Steiner drink wine?*

It seems that Rudolf Steiner did not drink wine. At least, this is what I have been told. Let us pick up here again the principle developed in the eighth lecture which suggests that one is what one eats, or rather, one is composed of what one digests (see question 31: What is the content of the Agriculture Course?). The result of this is that food has an influence on the physical as well as the psychological and even spiritual state of an individual, and thus a collective influence on social functioning. From this perspective, what is the specific role of alcohol? Steiner talks about it on various occasions, specifically in his lectures pertaining to Christianity (wine indeed occupies a central place in the Eucharist during mass). For him, alcohol contributed to a growing independence from the cosmos and to self-awareness. In this sense, its effect has been positive over the last few millennia as it helped mankind to develop greater autonomy of thinking. But today, the situation is different. Mankind has become so autonomous that we are disconnected, cut off from nature and the cosmos.

> In the course of evolution, it has had a mission. Strange as it may seem, it had the task, as it were, of preparing the human body so that it might be cut off from the Divine, in order to let the personal 'I am' emerge.

> Alcohol has the effect of severing the connection of the human being with the spirit world (*Gospel of St John,* p. 95, GA 103–092).

Or again, Steiner described how alcohol acts as a kind of interior imitation of the sun, affecting the 'I' *(From Comets,* pp. 216f). For this reason he believed that a person can more easily mobilise the forces of their 'I' if they abstain from alcohol. 'The one who seeks the initiation must attain the greatest peacefulness. This excludes the use of any stimulant, especially alcohol' *(Kosmogonie,* p. 203).

These are the ideas which explain why certain uncompromising biodynamists have looked down upon the application of biodynamics to vine-growing. I would like to make two remarks, however. First of all, this apparently posed no problem for Steiner who gave specific, practical advice on vine-growing (vine behaviour, issues surrounding phylloxera, and so on). Secondly, the grape has an extraordinary concentration of cosmic forces. And, in addition, Steiner explains the important role of the transformation brought about by fermentation, a process which concentrates even more fruit forces and makes them available to human beings. Wine could be an exceptional food in this sense ... Yet, in the end, it is a great irony of history that biodynamics is developing mostly in wine today, and can, therefore, reach the general public.

34 *What is the Goetheanum?*

The Goetheanum is the world centre for anthroposophy. It is an immense building designed by Rudolf Steiner himself who named it in honour of Goethe. It is situated on the top of a hill in Dornach, Switzerland, near Basel. The current building is in fact the second Goetheanum, built on the ruins of the first. Work on the first Goetheanum began in 1913 on land donated by a member of the Anthroposophic Society. It was inaugurated in September 1920. It was a large, wooden structure with a double dome and a 900-seat theatre. It was destroyed by arson on New Year's Eve, 1922. Steiner quickly worked on the construction of a second, even larger Goetheanum. He developed the plans in 1924, and its structure was built between 1925 and 1928 (the date of its official inauguration), and was completed after Steiner's death. At the time, it was truly a monument of expressionist architecture, and one of the very first large-scale buildings in Europe to be made of reinforced concrete. It houses two auditoria (1,000 and 450 seats, respectively), a library, offices, classrooms and conference rooms. Yet in fact, at the time of its inauguration in 1928, the building was incomplete and would remain so until 1970. Final works were not completed until 1998. In the Goetheanum, anthroposophic art is on show everywhere: architecture, sculptures, paintings, stained glass windows, and in theatre and dance performances.

Today, the Goetheanum is the headquarters of the anthroposophic movement, as well as the gathering place for conventions. Thus it is like a window into Steiner's heritage. It also houses the School of Spiritual Science, divided into various sections which explore each of the fields for which Rudolf Steiner urged renewal: medicine, agriculture, pedagogy, social sciences, and fine arts. The Goetheanum is therefore a venue for research and exchange. In a certain sense, it is the Steiner university campus, and it is possible to study at the Goetheanum throughout your life.

Should you have the opportunity to visit the Goetheanum, I would certainly recommend doing so. First of all, you will become aware of the vastness of Steiner's research and, before him, that of Goethe. Biodynamics is a very small part of it. Secondly, a visit will allow you to grasp Steiner's ideas through the atmosphere of architecture and the art. It is a valuable addition to reading his works.

35 *Viticulture after biodynamics: what next?*

This brief presentation of how biodynamics emerged enables you now to understand why this is a good method of cultivation, perhaps the best available today. It also allows us to rediscover what was lost with the development of modern agrochemistry: the connection to nature's subtle energies. Once this link is restored, it becomes possible to grow healthy, balanced, nutritious plants, while ceasing to destroy the soil and the environment. However, it is important to remember that biodynamics is not a *goal in itself*, but rather a *path*, or a way (see question 32: Are biodynamists a sect?). In this sense, I have noticed that once an individual engages in the practice of biodynamics, it is difficult to turn back. Very few growers stop practising biodynamics after several years, and even fewer stop when biodynamics is practised with conviction and not for commercial purposes. Indeed, what is discovered can no longer be denied, and that which is understood, is understood once and for all. I would even say that beyond a simple technique, the practice of biodynamics can also be lived by some as a path of personal development!

Here are some new areas that I see currently being developed by the most advanced biodynamic growers. These are often ancestral methods, recently rediscovered:

- *Goethean botany*. Rudolf Steiner gave new life to this old approach. It is founded on the meticulous observation of plants seeking to discern the forces which express themselves through it. By establishing a connection to the plant, one can, through a form of communication, perceive its imbalances and needs.
- *Geobiology*. This involves the study of a subsoil's impact on living organisms and, to a wider extent, all the telluric phenomena (the presence of bedrock, faults, underground water, telluric energy networks, and so on). This knowledge was used by the elders in ancient times for all sorts of constructions: residences and especially places of worship (temples, churches). And so why not by the monks who defined so precisely the different *terroirs* in Burgundy?
- *Bioenergy*. Originally developed with a therapeutic objective for people, its principle is to perform treatment on the energetic body before the diagnosed imbalance manifests itself as a pathology in the physical body. Plants also have an energetic body.
- *Feng shui*. Literally 'wind and water'. This Chinese science encompasses geobiology and focuses as much on natural spaces as on human structures, for which it is best known. Applied to the construction of wineries, it guarantees a harmonious atmosphere, favourable to the vinification and ageing of wine. Western architects had lost this ancient knowledge, and are now rediscovering it through this Eastern wisdom.
- *Informational techniques*. As with homeopathy, information can be transmitted by matter without any indication of a chemical or quantitative physical reaction. Water (and wine, which consists of 85% water) is particularly receptive to this. On this subject,

> I would recommend looking at the work carried out by a Japanese, Masaru Emoto, on the influence of thought on water structure. In the winery, this approach is very promising as a replacement for the arsenal of modern oenological products.

I could continue to cite several more, as there is no lack of related experiments being carried out. With the increasing spread of biodynamics, many rich and varied options are now available to explore along the way.

Conclusion

Dear reader, we have now reached the end of our journey together along the path of wine and biodynamics. I hope you will continue along the route yourself, equipped with the explanations gathered in this book. Yet, before parting, I would like to share with you a final thought.

You now understand the importance of the notion of *energy*, where lies the major difference between biodynamics and conventional agriculture. One takes it into account, the other ignores it completely. It is fascinating to observe that this split does not exist in Eastern culture where *chi* is still a reality accepted by all. Feng Shui talks about *chi* of the ground and *chi* of the sky. During a recent trip to Asia with my wife who is a Feng Shui practitioner, we had the opportunity to discuss this subject with an Asian tea expert. Her definition of Feng Shui for the cultivation of tea corresponded precisely with what we call *terroir* for wine! Is it therefore necessary to take into consideration a certain 'energy of place' in order to fully understand the idea of *terroir*? I recall a passionate discussion I had on this subject with writer and filmmaker, Georges Bardawil. We had been brought together by a mutual Japanese friend, Mr Kazuhiro Ota, during a dinner at the great Burgundian gastronomic restaurant Lameloise. For Georges Bardawil, 'Biodynamics is an excellent growing technique to promote the circulation of the good energies of *terroir*, and certainly the best available today.'

However, this 'energetic', and I would even say *spiritual,* vision of nature which the Orientals have preserved, is not reserved for them alone. In the past, it was profoundly anchored in our Western culture and needs only to radiate in a new way. It was powerfully present with the Benedictine and Cistercian monks who designated the Burgundian *terroirs.* I became aware of this during a tasting with an Orthodox monk by the name of Brother Jean. For this former international reporter and photographer, 'Nature is an open Book for those who cultivate the ground with their hands.' His religion: everyone can have a concrete, spiritual experience through simple acts in connection with nature: gardening, peeling a carrot ... (see *Le jardin de la foi,* published by Presses de la Renaissance).

Even in the West, this 'energetic' understanding of nature, which biodynamics seeks to formalise, is not reserved for monks or mystics. For those who are still sceptical, I will simply remind you of the words of Albert Einstein: 'Reality is merely an illusion, albeit a very persistent one.' It is up to each individual either to remain a prisoner of outdated ideas, or to free oneself from them. The only ideas one can be certain of are those that we have tested ourselves, advocated Descartes. Yet, this must be done without prejudice and in all intellectual honesty. As someone closer to us, the great American physicist Richard Feynman, observed, 'The first principle is to not deceive oneself; yet, you are the easiest person to deceive.'

In a time when agriculture is dominated by widespread agrochemical and genetically modified methods, the small voice of biodynamics seeks to make itself heard. Through wine, it touches both our senses and our inner soul.

References

Bardawil, Georges et Rozenbaum, Isabelle (2007) *Une promesse de vin,* Éditions Minerva.

Bouchet, François *Interview par Luc et Marie Boussard, Janvier 2005,* www.deuxversants.com.

Bouchet, François (2003) *Cinquante ans de pratique et d'enseignement de la bio-dynamie: son application dans la vigne,* Deux Versants Éditeur.

Bourguignon, Claude et Lydia (2008) *Le sol, la terre et les champs,* Éditions Sang de la Terre.

Étude Agreste 2006 *Enquête sur les pratiques culturales des viticulteurs,* Ministère de l'alimentation, de l'agriculture et de la pêche.

Frère Jean et Brobinskoy, Boris (2008) *Le jardin de la foi,* Presses de la Renaissance.

Joly, Nicolas (2005) *Wine from Sky to Earth,* Acres, USA.

Joly, Nicolas *Biodynamic wine demystified,* Wine Appreciation Guild.

Keyserling, Adalbert von *The Birth of a New Agriculture: Koberwitz 1924 and the Introduction of Biodynamics,* Temple Lodge Publishing.

Masson, Pierre (2011) *A Biodynamic Manual: Practical Instructions for Farmers and Gardeners,* Floris Books.

Masson, Pierre *Agenda Biodynamique Lunaire et Planétaire,* Biodynamie Services.

Rigaux, Jacky *La dégustation géo-sensorielle,* Éditions Terres en Vues.

Steiner, Rudolf *From Comets to Cocaine,* Rudolf Steiner Press, UK.

Steiner, Rudolf *The Gospel of St John,* Anthroposophic Press, USA.

Steiner, Rudolf *Kosmogonie,* Rudolf Steiner Verlag, Dornach (not translated into English).

Steiner, Rudolf *Spiritual Foundations for the Renewal of Agriculture,* Biodynamic Farming and Gardening Assoc., USA.

Steiner, Rudolf (1997) *An Outline of Esoteric Science,* Anthroposophic Press, USA.

Steiner, Rudolf *Spiritual Foundations for the Renewal of Agriculture,* Biodynamic Farming and Gardening Assoc., USA.

Steiner, Rudolf (1997) *An Outline of Esoteric Science,* Anthroposophic Press, USA.

Thun, Matthias *The Maria Thun Biodynamic Calendar,* Floris Books.

Thun, Matthias *When Wine Tastes Best: A Biodynamic Calendar for Wine Drinkers,* Floris Books.

Waldin, Monty (2006) *Biodynamic Wines,* Mitchell Beazley.

Waldin, Monty (2013) *Monty Waldin's Best Biodynamic Wines,* Floris Books.

Monty Waldin's Best Biodynamic Wines

'Bookshelves are not shy of wine buying guides, but Waldin's biodynamic selection is a valuable addition for any wine lover'.
— JANCISROBINSON.COM

'A thoroughly well done and useful publication.'
— TOM CANNAVAN, WINE JOURNALIST

Richly textured, vibrant chardonnay; mouthwateringly deep pinot gris; caressingly soft shiraz; opulent, forthright champagne. Monty Waldin, wine expert and star of Channel 4's 'Chateau Monty', shows us what is so wonderful about biodynamic wine in this comprehensive guide for wine drinkers.

Also available as an eBook

florisbooks.co.uk